Supporting open minds since 2005

Vortex Dynamics - From Physical to Mathematical Aspects
http://dx.doi.org/10.5772/intechopen.95649
Edited by İlkay Bakırtaş and Nalan Antar

Contributors
Daniel Duda, Heremba Bailung, Yoshiko Bailung, Takashi Suzuki, Ken Sawada, Hector Pérez-de-Tejada, Rickard Lundin, Kholoud Elmabruk, Sekip Dalgac, Naoto Ohmura, Hayato Masuda, Steven Wang, Rafael Bardera, Juan Carlos Matías, Estela Barroso, Stefania Espa, Maria Grazia Badas, Simon Cabanes, İlkay Bakırtaş, Mahmut Bağcı, Nalan Antar, Melis Turgut

Notice
Statements and opinions expressed in the chapters are these of the individual contributors and not necessarily those of the editors or publisher. No responsibility is accepted for the accuracy of information contained in the published chapters. The publisher assumes no responsibility for any damage or injury to persons or property arising out of the use of any materials, instructions, methods or ideas contained in the book.

First published in London, United Kingdom, 2022 by IntechOpen
IntechOpen is the global imprint of INTECHOPEN LIMITED, registered in England and Wales, registration number: 11086078, 5 Princes Gate Court, London, SW7 2QJ, United Kingdom
Printed in Croatia

British Library Cataloguing-in-Publication Data
A catalogue record for this book is available from the British Library

Additional hard and PDF copies can be obtained from orders@intechopen.com

Vortex Dynamics - From Physical to Mathematical Aspects
Edited by İlkay Bakırtaş and Nalan Antar
p. cm.
Print ISBN 978-1-80355-024-4
Online ISBN 978-1-80355-025-1
eBook (PDF) ISBN 978-1-80355-026-8

We are IntechOpen,
the world's leading publisher of
Open Access books
Built by scientists, for scientists

6,000+
Open access books available

146,000+
International authors and editors

185M+
Downloads

Our authors are among the

156
Countries delivered to

Top 1%
most cited scientists

12.2%
Contributors from top 500 universities

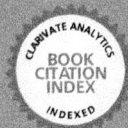

Interested in publishing with us?
Contact book.department@intechopen.com

Numbers displayed above are based on latest data collected.
For more information visit www.intechopen.com

Meet the editors

Dr. İlkay Bakırtaş is a Professor of Applied Mathematics in the Department of Mathematics, Istanbul Technical University (ITU), Turkey. She received her Ph.D. in Mechanics from ITU in 2003. She completed her postdoctoral studies at the University of Colorado in Boulder, USA. She has published eighteen research papers, four book chapters, and twenty-one conference proceedings in the fields of perturbation methods, nonlinear wave propagation in arteries, optical solitons and wave collapse in optics, and water waves problems. She has also edited two books. Dr. Bakırtaş is a member of the Scientific Committee of the Turkish National Committee of Theoretical and Applied Mechanics (TUMTMK). She received the 2004 Serhat Ozyar Young Scientist of The Year Award and the 2003 Best Ph.D. Dissertation Award from TUMTMK.

Dr. Nalan Antar is a Professor of Applied Mathematics in the Department of Mathematics Engineering, Istanbul Technical University (ITU), Turkey. She received her Ph.D. in Mechanics from ITU in 1999. She completed her post-doctoral studies at the Department of Mathematics and Statistics, University of Alberta, Canada, and later participated in academic research projects at the University of Colorado at Boulder, USA. She has published thirty-two research papers in peer-reviewed journals, thirteen conference proceedings, and three book chapters in the fields of nonlinear wave propagation in arteries, optical solitons in nonlinear optics, and water waves problems, in particular gravity currents. She has also edited two books. Dr. Antar has supervised many graduate students in applied mathematics. She is a member of the Scientific Committee of the Turkish National Committee of Theoretical and Applied Mechanics (TUMTM).

Contents

Preface

This book discusses recent advances in vortex dynamics. It includes nine chapters written by twenty-two authors from Turkey, Czech Republic, Japan, China, Italy, Mexico, Sweden, India, and Spain.

Chapter 1 investigates the existence and stability properties of dipole solitons in a nonlocal nonlinear medium with self-focusing and self-defocusing quantic nonlinear responses. The second chapter is a review chapter in which the authors examine the existing literature on the propagation of coherent or partially coherent vortex beams through a random medium. Chapter 3 presents an algorithm developed to detect individual vortices via direct fitting of the measured velocity field. Chapter 4 discusses the existing literature on vortex dynamics in complex fluids by considering Taylor vortex flow. The fifth chapter discusses rotating fluid flows affected by a β-effect and blood flow through a natural or artificial valve in the left ventricle. The next chapter is a theoretical work that studies a system of partial differential equations that is related to point vortices that appear in fluid dynamics. Chapter 7 discusses the features of vortex structures that are shown to exist in the plasma wake of Venus and the momentum transport phenomena of the vortex motion. Chapter 8 is a discussion of the existence of vortex structures in a dusty plasma medium. Finally, Chapter 9 analyzes and compares the horseshoe vortex and some other known models applied to biomimetic Micro Aerial Vehicles (MAVs).

This book is a useful resource for researchers, scientists, and postgraduate students in academia as well as industry.

Dr. İlkay Bakırtaş and Dr. Nalan Antar
Faculty of Science and Letters,
Mathematical Engineering Department,
Istanbul Technical University,
Istanbul, Turkey

Section 1

Vortex Structures in Nonlinear Optics

Dipole Solitons in a Nonlocal Nonlinear Medium with Self-Focusing and Self-Defocusing Quintic Nonlinear Responses

Mahmut Bağcı, Melis Turgut, Nalan Antar and İlkay Bakırtaş

Abstract

Stability dynamics of dipole solitons have been numerically investigated in a nonlocal nonlinear medium with self-focusing and self-defocusing quintic nonlinearity by the squared-operator method. It has been demonstrated that solitons can stay nonlinearly stable for a wide range of each parameter, and two nonlinearly stable regions have been found for dipole solitons in the gap domain. Moreover, it has been observed that instability of dipole solitons can be improved or suppressed by modification of the potential depth and strong anisotropy coefficient.

Keywords: dipole solitons, nonlinear response, nonlocal nonlinear medium, quintic nonlinearity

1. Introduction

Many phenomena in nature are modeled mathematically using nonlinear differential equations. Traveling wave solutions of nonlinear partial differential equations play a significant role in nonlinear wave propagation problems that are observed in various fields such as nonlinear optics, fluid dynamics, plasma physics, elastic media, and biology [1]. Some of the solutions to such nonlinear wave propagation problems are called solitons, which are localized wave solutions.

Optical solitons are formed because of the balance between the medium's diffraction and the self-phase modulation [2]. As a consequence of this, an optical field that does not change its shape occurs during propagation [3]. Recently, spatial solitons that can be used for optical switching and processing applications [4] have been extensively investigated in nonlinear optical systems with external optical lattices. There is a considerable amount of research about this subject in the literature. In 2003, Segev et al. experimentally observed spatial solitons in optically induced periodic potentials [5]. Fundamental and vortex solitons with real or complex lattices have been investigated in optical media with the cubic Kerr-type [6–10], the saturable [11], and competing nonlinearities [12]. Moreover, the existence of solitons has been observed in aperiodic or quasicrystal lattice structures [13–18] and the lattices that possess defects [19, 20] and dislocations [21, 22].

The dynamics of solitons are governed by nonlinear Schrödinger (NLS) type equations in optical media with nonlinearities and/or external potentials as in the

referred studies. Additionally, the cubic nonlinear NLS equation needs to be modified to describe nonlinear optical materials that have both cubic and quadratic
nonlinear responses [23–29], such as potassium niobate ($KNbO_3$) [30] or lithium
niobate ($LiNbO_3$) [31]. These dynamics in quadratically polarized media are
governed by the NLS equation with coupling to a mean term (d.c. field), which
are denoted as NLSM systems and sometimes referred to as Benney-Roskes or
Davey-Stewartson systems [32, 33].

NLSM equations were first studied by Benney and Roskes for water of finite
depth in the free surface conditions in 1969 [32]. Later, in 1974, Davey and
Stewartson derived the limiting integrable case, which is a reduced case of the
Benney–Roske's system by studying the evolution of a 3D wave packet for water of
finite depth [33]. In 1975, Ablowitz and Haberman [34] studied the integrability of
NLSM systems in the shallow water limit. The effects of surface tension were
included in the results of Benney and Roskes by Djordevic and Reddekopp [35] in
1977. From the first principles, Ablowitz et al. [23, 36, 37] discovered that NLSM-
type equations describe the evolution of the electromagnetic field in a quadratic
nonlinear media. The general NLSM system is given by [23, 36, 37]

$$iu_z + \Delta u + |u|^2 u - \rho u \phi = 0, \quad \phi_{xx} + v \phi_{yy} = \left(|u|^2 \right)_{xx}, \quad (1)$$

where $u(x, y, z)$ corresponds to the normalized amplitude of the envelope of the
static electric field propagating in the z direction, x and y are transverse spatial
coordinates. $\Delta u \equiv u_{xx} + u_{yy}$ corresponds to diffraction, the cubic term in u originates from the Kerr-type nonlinear change of the refractive index. The parameter ρ
is a coupling constant that comes from the combined optical rectification and
electro-optic effects modeled by the $\phi(x, y)$ field, and v is the coefficient that comes
from the anisotropy of the material [37]. Such systems of equations arise due to the
growth and depletion of the fundamental and second-harmonic fields at the
moment that the phase velocity of the fundamental and the second-harmonic wave
are not equal during propagation [38]. When the phase-matching condition is not
satisfied, the equation of the second-harmonic field can be solved directly and
generates an additional self-phase modulation contribution as a result of cascaded
nonlinearity. Similarly, the NLSM systems describe the nonlocal–nonlinear coupling
between the first harmonic with the cascading effect from the second harmonic and
a static field that is related to the mean term [36, 37].

Wave collapses play a significant role in various branches of science. The peak
amplitude of the wave solutions tends to infinity (blow-up) in finite time or finite
propagation distance when a singularity occurs. This phenomenon is often called
wave collapse [39]. In the NLS equation, it was first observed numerically by Kelley
in 1965 [40]. In fact, this wave collapse phenomenon is similar for the NLSM
systems. Wave collapse in the NLSM systems occurs with a modulated profile [41].
Merle and Raphael [42] analyzed the collapse behavior of the NLS equation and
other related equations in detail. Moreover, Moll et al. investigated experimental
observations of optical wave collapse in cubic nonlinearity and showed that the
amplitude of the wave increases as the spatial extent decreases in a self-similar
profile [43]. In Ref [39], Ablowitz and coworkers studied wave collapse that occurs
with a quasi-self-similar profile in the NLSM system and found that collapse can be
arrested by the small nonlinear saturation. Furthermore, in Ref [30], NLSM collapse
was arrested by wave self-rectification. In this aforementioned study, they considered only the nonlinear evolution of beams with an initial Gaussian beam profile
with several values of input power and/or beam ellipticity and found that the wave
collapse can be arrested by increasing the coupling constant ρ or for an initially
highly elliptic beam. Recently, the NLSM system collapse was arrested by adding a

real periodic [24] and partially parity-time-symmetric [44] and azimuthal [45] external lattices (potential) to the governing system, and it was shown numerically that modification of potential depth provides great controllability on the stability of soliton.

More recently, Bağcı et al. [46] have numerically investigated stability dynamics of fundamental lattice solitons that are solutions of extended NLSM system in a nonlocal nonlinear medium with self-focusing and self-defocusing quintic nonlinearity. It has been shown that as the absolute value of γ increases for both self-focusing and self-defocusing cases, the obtained fundamental solitons become nonlinearly unstable. However, the stability of unstable fundamental solitons can be improved by modification of potential depth [46].

Dipole (two-phased) and higher-phase vortex solitons in the presence of an induced lattice have been studied analytically and experimentally in Bose-Einstein condensates (BECs) [47, 48] and in optical Kerr media [49–54]. In recent years, these types of solitons have attracted considerable interest because of their unique features and potential applications [55].

In this chapter, we numerically study the existence and stability of dipole soliton solutions of the NLSM system in a nonlocal nonlinear medium with the self-defocusing quintic nonlinear response by adding an external lattice. In fact, this study is about the dynamics of dipole solitons instead of fundamental solitons in the problem that Bağcı and coworkers have addressed in recent book chapter [46]. The purpose of this study is to numerically investigate the effects of the strength of quintic nonlinearity that specify characteristics of the model and variation of potential depth on the existence and stability of dipole solitons. In several applications, many optical materials such as chalcogenide glasses are required quintic and seventh-order effects in addition to cubic nonlinear effects [56], and effective higher-order nonlinearities can reveal with pure Kerr materials in an inhomogeneous propagation media [57–59].

The chapter is outlined as follows: In Sec. 2, we present the model equations, and the squared-operator method is explained so that it is modified for the model. The dipole solitons are computed by this numerical method. Nonlinear evolution of the dipole solitons is examined to perform stability analysis, In Sec. 3. Finally in Sec. 4, results of this study are outlined.

2. The model

In this chapter, we modify the NLSM system (1) as follows to describe the dynamics of lattice solitons in a nonlocal nonlinear medium with cubic and quintic nonlinearity

$$iu_z + \frac{1}{2}\Delta u + \beta|u|^2 u - \rho u\phi + \gamma|u|^4 u - V(x,y)u = 0,$$

$$\phi_{xx} + v\phi_{yy} = \left(|u|^2\right)_{xx} \tag{2}$$

where γ is the coefficient of quintic nonlinearity and $V(x, y)$ is the optical lattice. In this chapter, we consider lattices that can be written as the intensity of a sum of N phase-modulated plane waves [13]

$$V(x,y) = \frac{V_0}{N^2}\left|\sum_{n=0}^{N-1} e^{i\left(k_x^n x + k_y^n y\right)}\right|^2, \tag{3}$$

where $V_0 > 0$ is the peak depth of the potential and the wave vector $\left(k_x^n, k_y^n\right) = [K \cos(2\pi n/N), K \sin(2\pi n/N)]$. The potential for $N = 2, 3, 4, 6$ yield crystal (periodic) lattices, while $N = 5, 7$ yield quasi-crystals (aperiodic) lattices. Contour image, contour plot, and diagonal cross-section of the lattice $V(x, y)$ are plotted in **Figure 1** for $V_0 = 12.5$, $N = 4$ and $k_x = k_y = 2\pi$. It can be seen that the lattice is periodic, and the center of lattice is a local maximum.

2.1 Numerical solution for the dipole solitons

Yang and Lakoba developed an iterative numerical method called the squared-operator method (SOM) [60]. The idea of this method is to iterate a modified differential equation whose linearization operator is square of the original equation together with a preconditioning (or acceleration) operator. To obtain the soliton solution of the (2+1)D NLSM model, this method is modified as follows:

Soliton solutions are sought in the form $u(x, y, z) = U(x, y)e^{i\mu z}$ where $U(x, y)$ is real-valued function and μ is the propagation constant (or eigenvalue). Substituting the ansatz $u(x, y, z)$, we get the following expressions:

$$u_z = i\mu U e^{i\mu z},$$

$$u_{xx} = U_{xx}e^{i\mu z},$$

$$u_{yy} = U_{yy}e^{i\mu z}, \tag{4}$$

$$|u|^2 = U e^{i\mu z} U^* e^{-i\mu z} = |U|^2,$$

$$|u|^4 = |U|^2|U|^2 = |U|^4$$

where $U = U^*$ in our case. Substituting the set of the terms in Eq. (4) into the (2+1) NLSM model, the following nonlinear equations for U are obtained

$$-\mu U + \frac{1}{2}\Delta U + \beta|U|^2 U - \rho\phi U + \gamma|U|^4 U - VU = 0,$$

$$\phi_{xx} + v\phi_{yy} = \left(|U|^2\right)_{xx}. \tag{5}$$

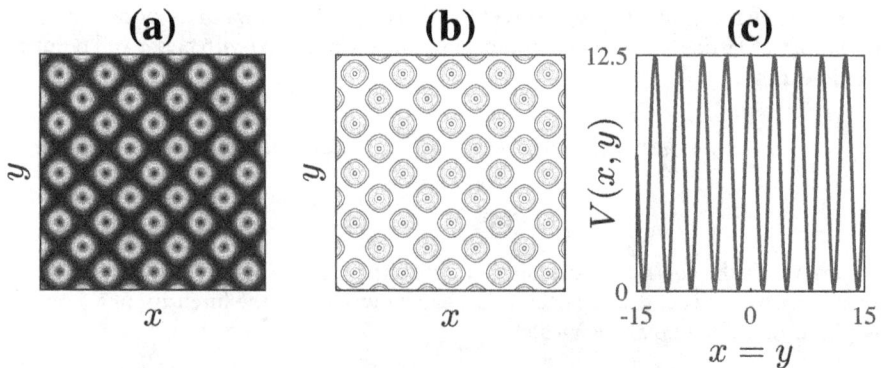

(a)   **(b)**   **(c)**

Figure 1.
(a) Contour image, (b) contour plot, and (c) diagonal cross-section of the lattice $V(x, y)$ when $V_o = 12.5$, $N = 4$ and (x,y) [15,15].

Applying the Fourier transform to the eigenequations system (5) yields

$$-\mu\hat{U} - \frac{1}{2}\left(k_x^2 + k_y^2\right)\hat{U} + \mathcal{F}\left\{\beta|U|^2U - \rho\phi U + \gamma|U|^4U - VU\right\} = 0,$$
$$k_x^2\hat{\phi} + vk_y^2\hat{\phi} = k_x^2\mathcal{F}\left\{|U|^2\right\},$$
(6)

where $\mathcal{F}$ denotes the Fourier transform, $\hat{U} = \mathcal{F}\{U\}$, k_x and k_y are the Fourier transform variables. Isolating $\hat{\phi}$ from the second equation of Eq. (6) gives

$$\hat{\phi} = \frac{k_x^2\mathcal{F}\left\{|U|^2\right\}}{k_x^2 + vk_y^2}.$$
(7)

Taking the inverse Fourier transform of Eq. (7), we get

$$\phi = \mathcal{F}^{-1}\left\{\frac{k_x^2\mathcal{F}\left\{|U|^2\right\}}{k_x^2 + vk_y^2}\right\},$$
(8)

where $\mathcal{F}^{-1}$ denotes the inverse Fourier transform and during iteration, the first element of $k_x^2 + vk_y^2$ is set to 1 in order to avoid division by zero error. By applying the inverse Fourier transform to first equation of Eq. (6) and substituting Eq. (8) into the obtained equation, we get

$$-\mu U + \mathcal{F}^{-1}\left\{-\frac{1}{2}\left(k_x^2 + k_y^2\right)\hat{U}\right\} + \beta|U|^2U - \rho\mathcal{F}^{-1}\left\{\frac{k_x^2\mathcal{F}\left\{|U|^2\right\}}{k_x^2 + vk_y^2}\right\}U + \gamma|U|^4U - VU = 0.$$
(9)

To obtain operator L_0, Eq. (9) can be written as

$$L_0U = \mathcal{F}^{-1}\left\{-\frac{1}{2}\left(k_x^2 + k_y^2\right)\hat{U}\right\} + T_0U = 0,$$
(10)

where

$$T_0 = -\mu + \beta|U|^2 - \rho\mathcal{F}^{-1}\left\{\frac{k_x^2\mathcal{F}\left\{|U|^2\right\}}{k_x^2 + vk_y^2}\right\} + \gamma|U|^4 - V.$$
(11)

Now, we should obtain operator L_1, which denotes the linearized operator of $L_0U = 0$, with respect to the solution U, i.e., $L_0(U + \tilde{U}) = L_1\tilde{U} + O(\tilde{U}^2)$, where $\tilde{U} \ll 1$. However, it should be noted that we have obtained the operator L_0 by substituting the mean field term $\phi(x, y)$ into the governing equation. Therefore, at this point, we have to perturb $\phi(x, y)$ function as well. In accordance with this purpose, the soliton solution and the mean-field term should be perturbed as follows, respectively,

$$u(x, y, z) = [U(x, y) + \tilde{U}(x, y)]e^{i\mu z},$$
$$\phi(x, y) = \phi(x, y) + \tilde{\phi}(x, y)$$
(12)

where $\tilde{U} \ll 1$ and $\tilde{\phi} \ll 1$. Firstly, consider that Eq. (5) can be written as a general type of nonlinearities

$$-\mu U + \frac{1}{2}\Delta U + F\left(|U|^2\right)U - \rho\phi U - VU = 0,$$

$$\phi_{xx} + v\phi_{yy} = \left(|U|^2\right)_{xx}, \tag{13}$$

where $F\left(|U|^2\right) = \beta|U|^2 + \gamma|U|^4$ for cubic-quintic nonlinearity. Then, substituting perturbation $U + \tilde{U}$ and using linear Taylor expansion yield

$$F\left(|U|^2\right) = F\left(|U + \tilde{U}|^2\right) = F\left(U^2 + 2U\tilde{U} + \tilde{U}^2\right)$$

$$\approx F(U^2 + 2U\tilde{U}) \tag{14}$$

$$\approx F\left(|U|^2\right) + 2U\tilde{U}F'_{|U|^2}\left(|U|^2\right)$$

where $U = U^*, \tilde{U} = \tilde{U}^*$ in our case and $F'_{|U|^2} = \partial F/\partial|U|^2$

$$F\left(|U|^2\right)U = F\left(|U + \tilde{U}|^2\right)(U + \tilde{U})$$

$$\approx \left[F\left(|U|^2\right) + 2U\tilde{U}F'_{|U|^2}\left(|U|^2\right)\right](U + \tilde{U})$$

$$\approx (U + \tilde{U})F\left(|U|^2\right) + 2U^2\tilde{U}F'_{|U|^2}\left(|U|^2\right) + 2U\tilde{U}^2F'_{|U|^2}\left(|U|^2\right) \tag{15}$$

$$\approx (U + \tilde{U})F\left(|U|^2\right) + 2U^2\tilde{U}F'_{|U|^2}\left(|U|^2\right) + O\left(\tilde{U}^2\right).$$

Substituting perturbations in Eq. (12) and Eq. (15) into Eq. (13) and only terms of $O(\tilde{U})$ and $O(\tilde{\phi})$ are retained, we get

$$-\mu(U + \tilde{U}) + \frac{1}{2}\Delta(U + \tilde{U}) + (U + \tilde{U})F\left(|U|^2\right) + 2U^2\tilde{U}F'_{|U|^2}\left(|U|^2\right)$$

$$-\rho\phi(U + \tilde{U}) - \rho\tilde{\phi}U - V(U + \tilde{U}) = 0, \tag{16}$$

$$\phi_{xx} + v\phi_{yy} + \tilde{\phi}_{xx} + v\tilde{\phi}_{yy} = \left(|U|^2\right)_{xx} + (2U\tilde{U})_{xx}.$$

Substituting $F\left(|U|^2\right) = \beta|U|^2 + \gamma|U|^4 \Rightarrow F'_{|U|^2}\left(|U|^2\right) = \beta + 2\gamma|U|^2$ into Eq. (16) yields

$$-\mu(U + \tilde{U}) + \frac{1}{2}\Delta(U + \tilde{U}) + (U + \tilde{U})\left(\beta|U|^2 + \gamma|U|^4\right) + 2U^2\tilde{U}\left(\beta + 2\gamma|U|^2\right)$$

$$-\rho\phi(U + \tilde{U}) - \rho\tilde{\phi}U - V(U + \tilde{U}) = 0,$$

$$\phi_{xx} + v\phi_{yy} + \tilde{\phi}_{xx} + v\tilde{\phi}_{yy} = \left(|U|^2\right)_{xx} + (2U\tilde{U})_{xx}.$$

$$\tag{17}$$

Applying the Fourier transform to the first perturbed equation in Eq. (17) and the inverse Fourier transform to the obtained equation, we get

$$-\mu(U + \tilde{U}) + \mathcal{F}^{-1}\left\{-\frac{1}{2}\left(k_x^2 + k_y^2\right)\left(\hat{U} + \hat{\tilde{U}}\right)\right\} + (U + \tilde{U})\left(\beta|U|^2 + \gamma|U|^4\right)$$

$$+2U^2\tilde{U}\left(\beta + 2\gamma|U|^2\right) - \rho\phi(U + \tilde{U}) - \rho\tilde{\phi}U - V(U + \tilde{U}) = 0. \tag{18}$$

Using Eq. (5) and the second equation in Eq. (17), following equation is obtained

$$\tilde{\phi}_{xx} + v\tilde{\phi}_{yy} = (2U\tilde{U})_{xx}. \tag{19}$$

Applying Fourier transform to Eq. (19) and isolating $\hat{\tilde{\phi}}$ from obtained equation, we get

$$k_x^2\hat{\tilde{\phi}} + vk_y^2\hat{\tilde{\phi}} = k_x^2\mathcal{F}\{2U\tilde{U}\}$$

$$\hat{\tilde{\phi}} = \frac{k_x^2\mathcal{F}\{2U\tilde{U}\}}{k_x^2 + vk_y^2}. \tag{20}$$

Taking the inverse Fourier transform of $\hat{\tilde{\phi}}$ yields

$$\tilde{\phi} = \mathcal{F}^{-1}\left\{\frac{k_x^2\mathcal{F}\{2U\tilde{U}\}}{k_x^2 + vk_y^2}\right\}. \tag{21}$$

Substituting Eq. (8) and Eq. (21) into Eq. (18) yields

$$-\mu(U + \tilde{U}) + \mathcal{F}^{-1}\left\{-\frac{1}{2}\left(k_x^2 + k_y^2\right)\left(\hat{U} + \hat{\tilde{U}}\right)\right\} + (U + \tilde{U})\left(\beta|U|^2 + \gamma|U|^4\right)$$

$$+2U^2\tilde{U}\left(\beta + 2\gamma|U|^2\right) - \rho\mathcal{F}^{-1}\left\{\frac{k_x^2\mathcal{F}\{|U|^2\}}{k_x^2 + vk_y^2}\right\}(U + \tilde{U}) \tag{22}$$

$$-\rho\mathcal{F}^{-1}\left\{\frac{k_x^2\mathcal{F}\{2U\tilde{U}\}}{k_x^2 + vk_y^2}\right\}U - V(U + \tilde{U}) = 0.$$

After grouping the terms, Eq. (22) can be written as

$$\left[-\mu U + \mathcal{F}^{-1}\left\{-\frac{1}{2}\left(k_x^2 + k_y^2\right)\hat{U}\right\} + \beta|U|^2U - \rho\mathcal{F}^{-1}\left\{\frac{k_x^2\mathcal{F}\{|U|^2\}}{k_x^2 + vk_y^2}\right\}U + \gamma|U|^4U - VU\right]$$

$$+\left[-\mu\tilde{U} + \mathcal{F}^{-1}\left\{-\frac{1}{2}\left(k_x^2 + k_y^2\right)\hat{\tilde{U}}\right\} + \beta|U|^2\tilde{U} + \gamma|U|^4\tilde{U} + 2\beta|U|^2\tilde{U} + 4\gamma|U|^4\tilde{U}\right.$$

$$\left.-\rho\mathcal{F}^{-1}\left\{\frac{k_x^2\mathcal{F}\{|U|^2\}}{k_x^2 + vk_y^2}\right\}\tilde{U} - \rho\mathcal{F}^{-1}\left\{\frac{k_x^2\mathcal{F}\{2U\tilde{U}\}}{k_x^2 + vk_y^2}\right\}U - V\tilde{U}\right] = 0. \tag{23}$$

Hence,

$$\left[-\mu U + \mathcal{F}^{-1}\left\{-\frac{1}{2}\left(k_x^2 + k_y^2\right)\hat{U}\right\} + \beta|U|^2 U - \rho\mathcal{F}^{-1}\left\{\frac{k_x^2\mathcal{F}\{|U|^2\}}{k_x^2 + vk_y^2}\right\}U + \gamma|U|^4 U - VU\right]$$

$$+\left[-\mu\tilde{U} + \mathcal{F}^{-1}\left\{-\frac{1}{2}\left(k_x^2 + k_y^2\right)\hat{\tilde{U}}\right\} + 3\beta|U|^2\tilde{U} - \rho\mathcal{F}^{-1}\left\{\frac{k_x^2\mathcal{F}\{|U|^2\}}{k_x^2 + vk_y^2}\right\}\tilde{U} + 5\gamma|U|^4\tilde{U}\right.$$

$$\left.-\rho\mathcal{F}^{-1}\left\{\frac{k_x^2\mathcal{F}\{2U\tilde{U}\}}{k_x^2 + vk_y^2}\right\}U - V\tilde{U}\right] = 0.$$

$$(24)$$

From Eq. (9), we know that the first bracket is identically zero. Consequently, we obtain

$$-\mu\tilde{U} + \mathcal{F}^{-1}\left\{-\frac{1}{2}\left(k_x^2 + k_y^2\right)\hat{\tilde{U}}\right\} + 3\beta|U|^2\tilde{U} - \rho\mathcal{F}^{-1}\left\{\frac{k_x^2\mathcal{F}\{|U|^2\}}{k_x^2 + vk_y^2}\right\}\tilde{U} + 5\gamma|U|^4\tilde{U}$$

$$-\rho\mathcal{F}^{-1}\left\{\frac{k_x^2\mathcal{F}\{2U\tilde{U}\}}{k_x^2 + vk_y^2}\right\}U - V\tilde{U} = 0.$$

$$(25)$$

Moreover, Eq. (24) satisfied $L_0(U + \tilde{U}) = L_1\tilde{U} + O\left(\tilde{U}^2\right)$. Therefore, to obtain a linearized operator L_1L_1, Eq. (25) can be written as

$$L_1\tilde{U} = \mathcal{F}^{-1}\left\{-\frac{1}{2}\left(k_x^2 + k_y^2\right)\hat{\tilde{U}}\right\} + T_1\tilde{U} - \rho\mathcal{F}^{-1}\left\{\frac{k_x^2\mathcal{F}\{2U\tilde{U}\}}{k_x^2 + vk_y^2}\right\}U = 0, \quad (26)$$

where

$$T_1 = -\mu + 3\beta|U|^2 - \rho\mathcal{F}^{-1}\left\{\frac{k_x^2\mathcal{F}\{|U|^2\}}{k_x^2 + vk_y^2}\right\} + 5\gamma|U|^4 - V. \quad (27)$$

To obtain soliton solution $U(x, y)$ in $L_0U = 0$, we numerically integrate the following distance-dependent squared-operator evolution equation

$$U_z = -M^{-1}L_1^\dagger M^{-1}L_0U, \quad (28)$$

where $(\cdot)^\dagger$ denotes the Hermitian of the operator and M is a real-valued positive definite Hermitian preconditioning operator that is introduced to accelerate the convergence. Since it is easily invertible to take the Fourier transform, we take the preconditioning operator M to be in the form of the following

$$M = c - \left(\partial_{xx} + \partial_{yy}\right), \quad (29)$$

where $c > 0$ is a parameter for parametrizing the numerical scheme. Applying the Fourier transform to Eq. (29) yields

$$\mathcal{F}\{M\} = c + k_x^2 + k_y^2. \tag{30}$$

Consequently, in Eq. (28)

$$M^{-1}L_1^\dagger M^{-1}L_0 U = \mathcal{F}^{-1}\left\{\frac{\mathcal{F}\{L_1 M^{-1}L_0 U\}}{c + k_x^2 + k_y^2}\right\}. \tag{31}$$

Using the forward Euler method, steady-state solution U is computed by an iterative scheme as follows

$$U_{n+1} = U_n - \left[M^{-1}L_1^\dagger M^{-1}L_0 U\right]_{U=U_n} \Delta z, \tag{32}$$

where Δz is an auxiliary distance-step parameter. It has been demonstrated that the SOM algorithm converges to a soliton solution for a wide range of nonlinear PDEs if the initial condition is sufficiently close to the exact solution and the distance-step Δz in the iteration scheme is less than a specific threshold value [60, 61]. To obtain a convergent soliton solution, c and Δz are chosen heuristically as positive real numbers. Moreover, our convergence criterion is that the obtained solution satisfies Eq. (10) with an absolute error less than 10^{-5}.

In this chapter, to obtain dipole solitons, the initial condition of the SOM algorithm is chosen as a multi-humped Gaussian function which is given by

$$U_0(x, y, 0) = \sum_{n=0}^{H-1} e^{-A\left[(x+x_n)^2 + (y+y_n)^2\right] + i\theta_n}, \tag{33}$$

where x_n and y_n represent the location of the solitons on the lattice, H corresponds to the number of humps, A is a positive integer, and θ_n is the phase difference. Since we numerically investigate the dipole solitons, H is set to 2, thus Eq. (33) takes the following form:

$$U_0(x, y, 0) = e^{-A\left[(x+x_0)^2 + (y+y_0)^2\right] + i\theta_0} + e^{-A\left[(x+x_1)^2 + (y+y_1)^2\right] + i\theta_1}, \tag{34}$$

where (x_0, y_0) and (x_1, y_1) represent the locations of dipole solitons, θ_0 and θ_1 are the phase differences of dipole solitons. It was shown that the solitons located at the maximum of the lattices are unstable [13, 21, 24], due to this fact we will investigate the dipole solitons located on minima of the considered square lattice. A dipole (two-phased) localized soliton numerically found by

$$A = 1, \quad x_n = r\cos\theta_n, \quad y_n = r\sin\theta_n, \quad n = 0,1. \tag{35}$$

Here r is set to be π and $\theta_n = n\pi$, so that the humps of the initial condition are located at the local minima of the lattice where $(x_0, y_0) = (\pi, 0)$ and $(x_1, y_1) = (-\pi, 0)$.

Unless otherwise stated, parameters in the NLSM model (2) are fixed to

$$(\mu, \rho, v, \beta, \gamma, V_0) = (-0.1, 0.5, 1.5, 2, -0.1, 12.5). \tag{36}$$

It is noted that $\rho = 0.5$ and $v = 1.5$ are especially chosen to simulate quadratic optical effects in potassium niobate (KNbO$_3$) [30].

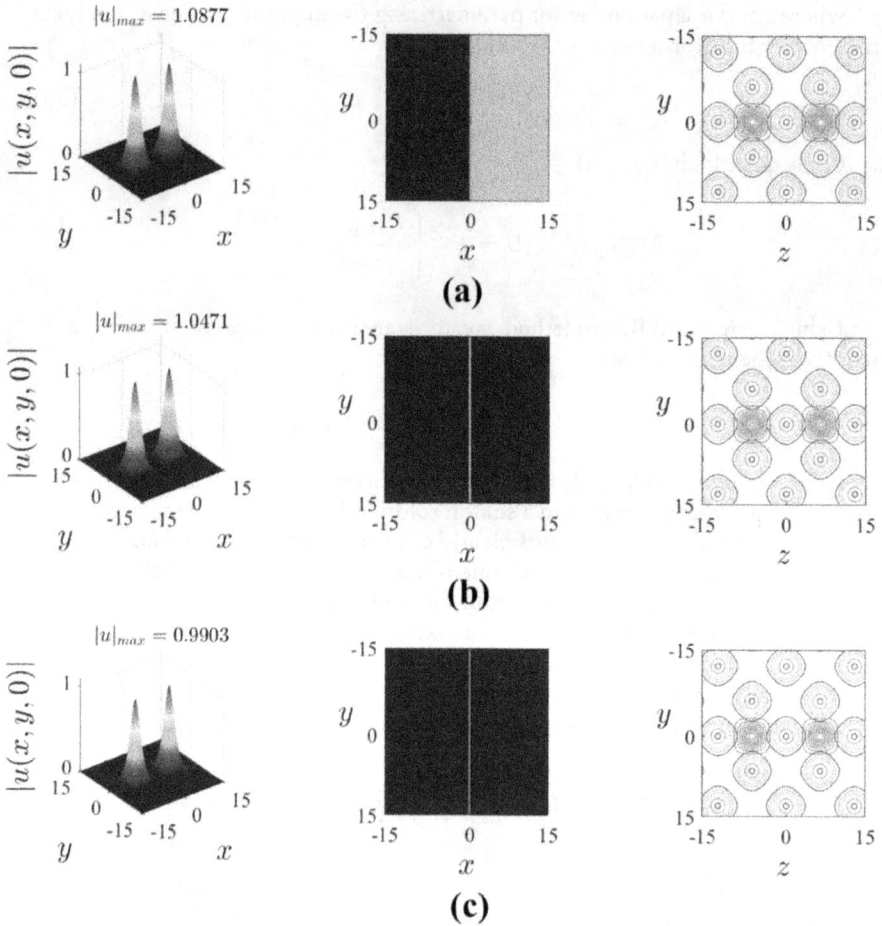

Figure 2.
3D dipole profiles centered at the lattice minima (first column), the phase structures of the dipole (second column), and the contour plot of the dipole solitons superimposed on the underlying lattice (third column), which are obtained for (a) $\gamma = -0.3, c = 2.1, \Delta z = 0.4$ and error is order of 10^{-6}, (b) $\gamma = -0.1, c = 2.5, \Delta z = 0.4$ and error is order of 10^{-8}, and (c) $\gamma = 0.3, c = 2.5, \Delta z = 0.4$ and error is order of 10^{-8}. All other parameters are fixed to the values in Eq. (36).

Dipole solitons of the NLSM model (2) are calculated by the SOM method. In **Figure 2**, 3D views (first column), phase structures (second column), and contour plots of dipole solitons on the underlying lattice (third column) are displayed for self-defocusing ($\gamma < 0$) and self-focusing ($\gamma > 0$) quintic nonlinearities. γ is set to be -0.3, -0.1 and 0.3 in **Figure 2(a)–(c)**, respectively, and all other parameters are fixed to the values given in Eq. (36). **Figure 2** shows that the dipole solitons can be generated on the lattice minima (see the third column), and the amplitudes of dipole solitons are decreased as γ increased (from -0.3 to 0.3) (see the first column).

3. Stability analysis

The stability dynamics of dipole solitons obtained by the SOM method are studied by the power analysis and direct simulation of the nonlinear evolution.
The power of solitons plays an important role in the stability analysis and it is calculated by

$$P(\mu) = \int_{-\infty}^{+\infty}\int_{-\infty}^{+\infty} |U(x, y; \mu)|^2 dxdy. \tag{37}$$

Vakhitov and Kolokolov proved a necessary condition for the linear stability of solitons in Ref [62]. They demonstrated that a soliton is linearly stable only if its power increases as propagation constant (or eigenvalue) μ increases. In other words, a necessary condition for the stability of solitons is

$$\frac{dP}{d\mu} > 0. \tag{38}$$

Moreover, Weinstein and Rose [63, 64] proved that a necessary condition for the nonlinear stability of solitons is also the slope condition given in Eq. (38).

To analyze nonlinear stability of the NLSM model (2), we examine the direct simulation of dipole solitons obtained by the SOM method. A finite-difference discretization scheme is used in the spatial domain (x, y) and the dipole solitons are advanced in the z direction with a fourth-order Runge-Kutta method. The initial condition of the nonlinear evolution is taken to be a dipole soliton, and 1% random noise is inserted into the amplitude of the initial condition.

The power diagrams of dipole solitons are displayed for varied μ, γ, β and ρ values in **Figure 3(a)–(d)**, respectively. It is noted that the domain of existence for the varied parameter is shown on the x-axis of each panel when other parameters are fixed to the values in Eq. (36). **Figure 3** shows that the power of dipole solitons increases as μ and ρ increase, whereas the power of dipole solitons decreases as γ and β increase. Moreover, the stability (solid blue) and instability (red dotted) regions of parameters are determined by the nonlinear evolution of dipole solitons for each point on the power curves.

The dipole solitons are found to be nonlinearly stable for self-defocusing quintic nonlinearity ($\gamma = -0.1$) when the power $P \in [0.99, 1.87]$ and propagation constant $\mu \in [-0.75, -0.6]$. Also, the dipole solitons are nonlinearly stable when $P \in [3.12, 4.26]$ and $\mu \in [-0.35, -0.06]$, which is the second nonlinearly stable gap (see **Figure 3(a)**). These results are consistent with key analytical results on nonlinear stability, which Weinstein and Rose proved in Ref [63, 64], since slope of the power-eigenvalue ($P - \mu$) diagram is positive. As can be seen from **Figure 3(b)**, the dipole solitons are obtained for $\gamma \in [-0.7, 25]$, when other parameters are fixed, and dipole solitons are nonlinearly stable for $\gamma \in [-0.21, 0.25]$. Zoom-in view of this stability domain is depicted in **Figure 3(b)**. Furthermore, it is observed that dipole solitons are nonlinearly stable for $\beta \in [1.6, 18.9]$ (see **Figure 3(c)**), and dipole solitons are stable for $\rho \in [0, 0.8]$ (see **Figure 3(d)**) in their existence domains when other parameters are fixed.

In **Figure 4**, nonlinear evolution of peak amplitudes, 3D views of the evolved dipole solitons, and the phase structures of evolved dipole solitons are plotted for the dipole solitons that are shown in **Figure 2**. The effect of quintic nonlinearity (γ) on nonlinear stability is investigated by fixing other parameters as in Eq. (36).

Figure 4(b) shows that peak amplitudes of dipole solitons oscillate mildly (first column), and the 3D profile (second column) and phase structure (third column) of dipole solitons are preserved for $\gamma = -0.1$. Thus, the dipole solitons are nonlinearly stable for the defocusing quintic nonlinearity for the considered parameter regime. On the other hand, as shown in **Figure 4(a)** and **(c)**, when the quintic nonlinearity is strong ($\gamma = -0.3$ and $\gamma = +0.3$), peak amplitudes of dipole solitons increase significantly in a short propagation distance z (first column), dipole profiles (second column) cannot be preserved, and phase structures of dipole solitons (third column) break up after evolution. Comparing **Figure 4(a)** and **(c)**, it is observed

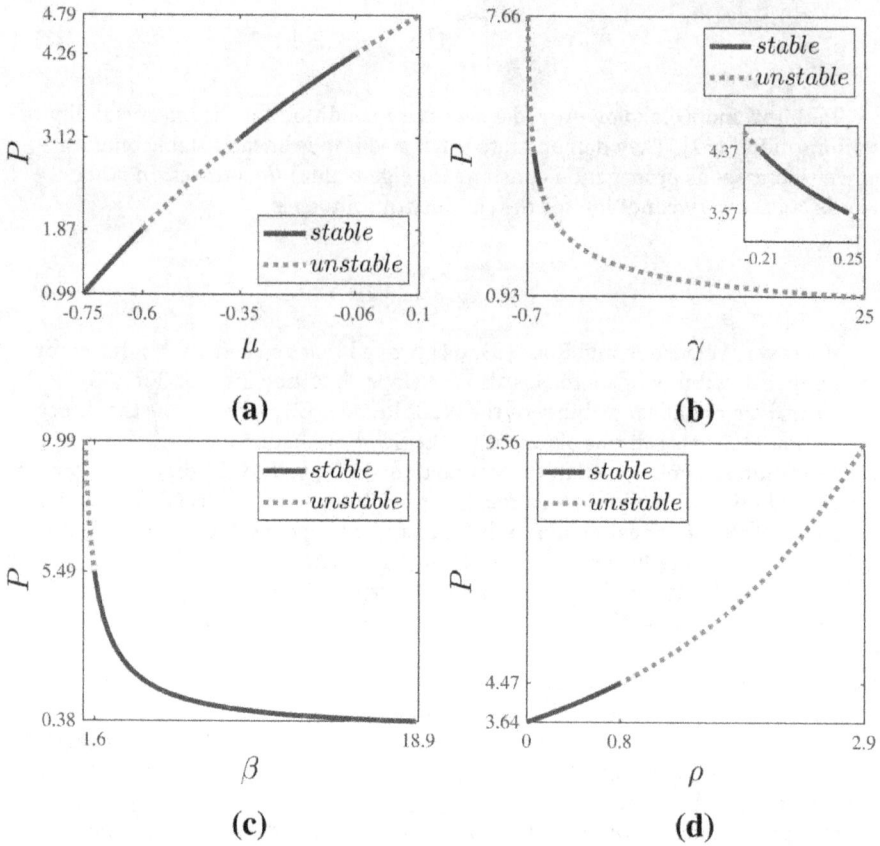

Figure 3.
Power of dipole solitons (a) for varying eigenvalue μ, (b) for varying quintic nonlinearity coefficient γ, (c) for varying cubic nonlinearity coefficient β, and (d) for varying quadratic nonlinear response ρ. The nonlinear stability and instability regions are shown by solid blue and red dotted lines, respectively.

that the propagation distance of dipole solitons in a medium with strong self-focusing nonlinearity ($\gamma = 0.3$) is longer than that of a medium with strong self-defocusing nonlinearity ($\gamma = -0.3$). Considering these evolution results in **Figure 4** and the existing domain for γ in **Figure 3(b)**, it is demonstrated that both strong self-focusing and self-defocusing quintic nonlinearities have a negative effect on the nonlinear stability of dipole solitons.

In previous studies [6, 14, 24], it is found that modification of the depth of potential can suppress nonlinear instabilities. More recently, Bağcı and coworkers [46] have demonstrated that nonlinear stability of fundamental solitons in an NLSM system (2) with quintic nonlinearity can be improved by the modification of lattice depth V_0. They showed that increased lattice depth supports the stability of fundamental solitons in a medium with strong self-focusing ($\gamma = 0.3$) quintic nonlinearity, and the stability of solitons in a medium with strong self-defocusing ($\gamma = -0.3$) quintic nonlinearity can be improved by decreasing lattice depth. For the dipole solitons, evolution of peak amplitudes is depicted for varying potential depths, when $\gamma = -0.3$ and $\gamma = 0.3$ in **Figure 5(a)** and **(b)**, respectively. **Figure 5 (a)** shows that the stability of dipole solitons is improved by decreasing lattice depth (from 25 to 5) for strong self-defocusing nonlinearity ($\gamma = -0.3$), and collapse can be arrested when $V_0 = 5$. In contrast, as shown in **Figure 5(b)**, the propagation distance of dipole solitons in a medium with strong self-focusing

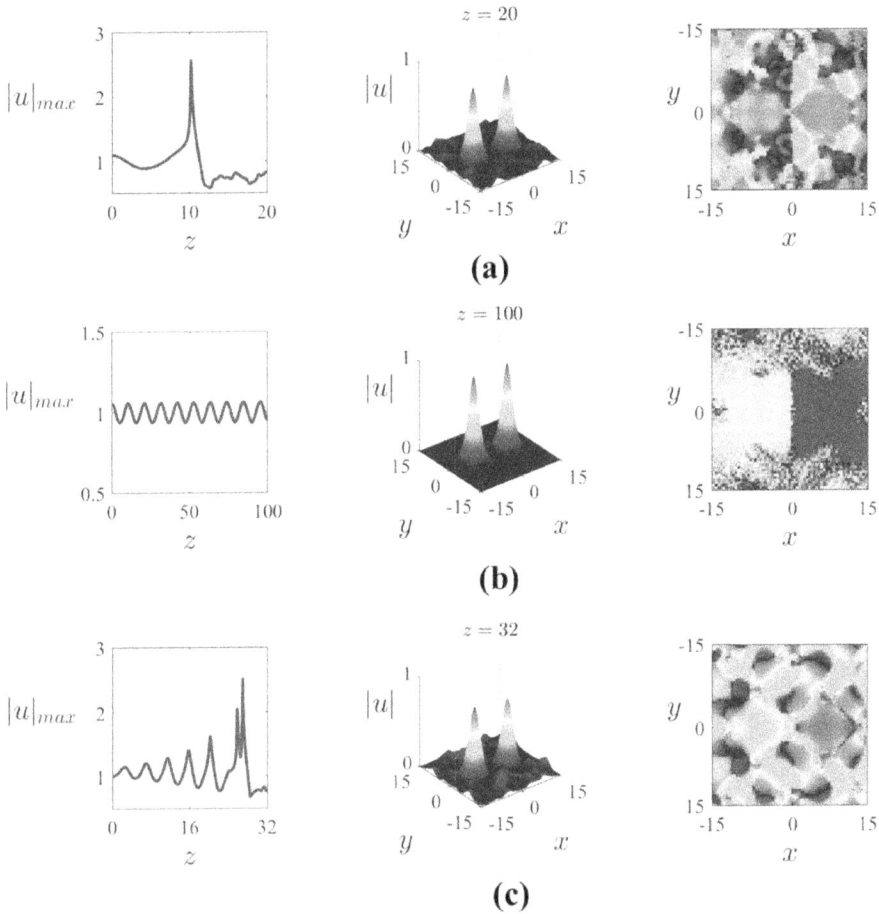

Figure 4.
Nonlinear evolution of maximum amplitudes as a function of propagation distance z (first column), 3D views of the dipole solitons after evolution (second column), and the phase structures of dipole solitons after evolution (third column) for (a) $\gamma = -0.3$, (b) $\gamma = -0.1$, and (c) $\gamma = 0.3$. All other parameters are taken as in Eq. (36).

nonlinearity ($\gamma = 0.3$) is extended by increasing lattice depth (from 5 to 50). It should be noted that these results are in agreement with the findings of the afore-mentioned studies. Thus modification of the lattice depth can be utilized to improve the nonlinear stability of dipole solitons.

It is also known that when the quadratic [15, 24, 44] and quintic [46] electro-optic effects are strong, the instability of fundamental solitons can be improved by increasing the anisotropy parameter. To examine the effect of anisotropy coefficient v on the nonlinear stability of dipole solitons in a medium with strong quintic nonlinearity ($\gamma = -0.3$ and $\gamma = 0.3$), evolution of the peak amplitudes is displayed for varied v values in **Figure 6**. **Figure 6** shows that increasing the anisotropy coefficient v from 0.001 to 10 stabilizes the dipole solitons in a medium with strong self-defocusing nonlinearity ($\gamma = -0.3$), and increasing v from 0.001 to 1000 extends the propagation distance of dipole solitons in a medium with strong self-focusing nonlinearity ($\gamma = 0.3$). Thus, larger anisotropy coefficient supports the nonlinear stability of the dipole solitons, and this result complies with the results of the previous studies [46]. It is important to note that the parameters ρ and v are predetermined coefficients that depend on the type of optical materials; larger

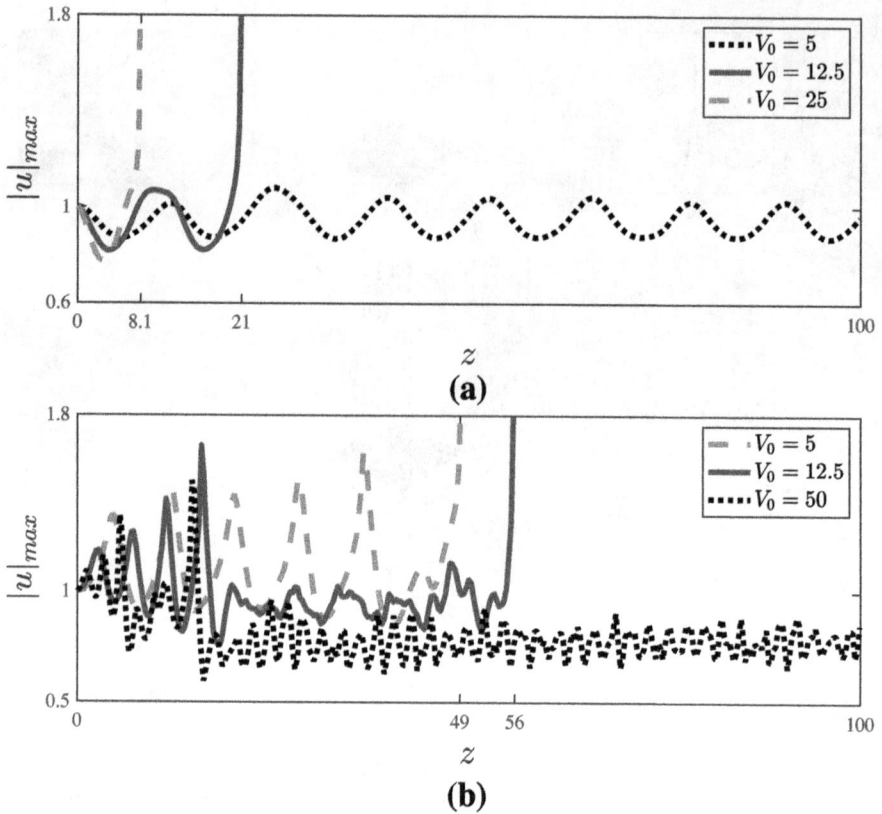

Figure 5.
Maximum amplitudes of the evolved dipole solitons for varying depth of potential V_o, when the dipole soliton is obtained for (a) $\gamma = -0.3$ and (b) $\gamma = 0.3$.

values of v cannot be applied to real optical systems. In this chapter, the effect of extremely large v values on the stability of dipole solitons is explored numerically.

4. Conclusions

In this chapter, the existence and nonlinear stability dynamics of dipole solitons have been investigated for a nonlocal nonlinear medium with quintic nonlinear response. This medium was characterized by the (2+1)D NLSM system with a periodic external lattice. Dipole solitons were obtained for self-defocusing ($\gamma < 0$) and self-focusing ($\gamma > 0$) quintic nonlinearities by the SOM method, and the nonlinear stability of these dipole structures has been investigated by the direct simulation of the model equations. Power of dipole solitons was determined for varying μ, γ, β, and ρ parameters and it has shown that the power of dipole solitons increases as the eigenvalue μ and quadratic nonlinear response ρ increase, whereas the power of dipole solitons decreases as quintic nonlinearity coefficient γ and cubic nonlinearity coefficient β increase.

Nonlinear evolution of the dipole solitons showed that the dipole solitons are stable for the weak self-focusing and self-defocusing quintic nonlinearity. In other words, as an absolute value of γ increases, the obtained dipole solitons become nonlinearly unstable in both self-focusing and self-defocusing media. It has been demonstrated that the collapse of dipole solitons can be arrested by decreased

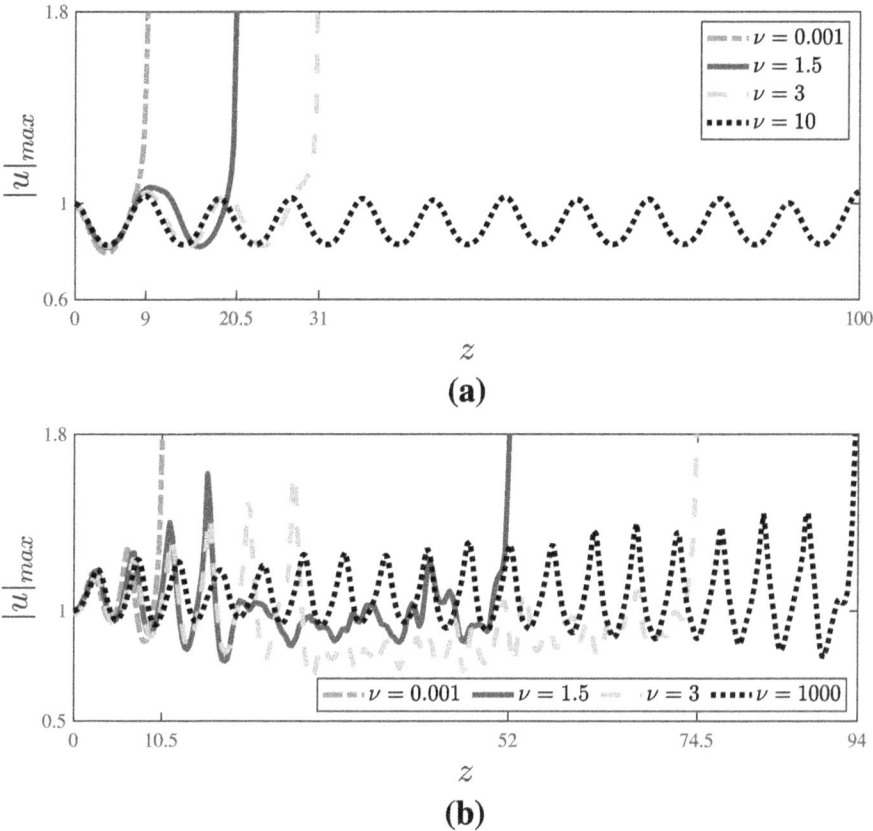

Figure 6.
Maximum amplitudes of the evolved dipole solitons for varying anisotropy coefficients v, when the dipole soliton is obtained for (a) γ = −0.3, and (b) γ = 0.3.

potential depth in a medium with strong self-defocusing quintic nonlinearity ($\gamma = -0.3$), while the deeper lattice extends the propagation distance of dipole solitons in a medium with strong self-focusing quintic nonlinearity ($\gamma = 0.3$). Furthermore, it has been observed that increasing the anisotropy coefficient (v) extends the propagation distance of the dipole solitons for strong self-focusing quintic nonlinearity, and it stabilizes the dipole solitons for strong self-defocusing quintic nonlinearity.

In conclusion, the existence and stability properties of dipole solitons have been numerically explored in a nonlocal nonlinear medium with quintic nonlinear response, and it has been demonstrated that the instability of dipole solitons can be suppressed by modification of the lattice depth and increased anisotropy coefficient.

Conflict of interest

The authors declare no conflict of interest.

Author details

Mahmut Bağcı[1*†], Melis Turgut[2†], Nalan Antar[2†] and İlkay Bakırtaş[2†]

1 Department of Management Information Systems, Marmara University, Istanbul, Türkiye

2 Department of Mathematical Engineering, Istanbul Technical University, Istanbul, Türkiye

*Address all correspondence to: bagcimahmut@gmail.com

† These authors contributed equally.

IntechOpen

References

[1] Ablowitz MJ. Nonlinear Dispersive Waves: Asymptotic Analysis and Solutions. Cambridge University Press; 2011

[2] Stegeman GIA, Christodoulides DN, Segev M. Optical spatial solitons: historical perspectives. IEEE Journal of Selected Topics in Quantum Electronics. 2000;**6**(6):1419-1427. DOI: 10.1109/2944.902197

[3] Yuri S, Agrawal GP. Optical Solitons: From Fibers to Photonic Crystals. Academic Press; 2003

[4] Andersen DR, Allan GR, Skinner SR, Smirl AL. Observation of fundamental dark spatial solitons in semiconductors using picosecond pulses. Optics Letters. 1991;**16**(3):156-158. DOI: 10.1364/OL.16.000156

[5] Fleischer JW, Segev M, Efremidis NK, Christodoulides DN. Observation of two-dimensional discrete solitons in optically induced nonlinear photonic lattices. Nature. 2003;**422**:147

[6] Ablowitz MJ, Antar N, Bakırtaş İ, Ilan B. Band-gap boundaries and fundamental solitons in complex two-dimensional nonlinear lattices. Physical Review A. 2010;**81**:033834

[7] Bağcı M. Impact of the lattice period on the stability dynamics of defect solitons in periodic lattices. Physical Review A. 2022;**105**:043524

[8] Christodoulides DN, Yang J. Parity-Time Symmetry and Its Applications. Singapore: Springer; 2018

[9] Yang J, Musslimani ZH. Fundamental and vortex solitons in a two-dimensional optical lattice. Optics Letters. 2003;**28**(21):2094-2096. DOI: 10.1364/OL.28.002094

[10] Kartashov LTYV, Egorov AA, Christodoulides DN. Stable soliton complexes in two-dimensional photonic lattices. Optics Letters. 2004;**29**(16):1918-1920. DOI: 10.1364/OL.29.001918

[11] Göksel İ, Bakırtaş İ, Antar N. Nonlinear lattice solitons in saturable media. Applied Mathematics and Information Sciences. 2014;**9**:377-385

[12] Göksel İ, Antar N, Bakırtaş İ. Two-dimensional solitons in PT-symmetric optical media with competing nonlinearity. Optik. 2018;**156**:470-478

[13] Mark J, Ilan B, Schonbrun E, Piestun R. Solitons in two-dimensional lattices possessing defects, dislocations, and quasicrystal structures. Physical Review E. 2006;**74**:035601

[14] Ablowitz MJ, Antar N, Bakırtaş İ, Ilan B. Vortex and dipole solitons in complex two-dimensional nonlinear lattices. Physical Review A. 2012;**86**: 033804

[15] Bağcı M. Soliton dynamics in quadratic nonlinear media with two-dimensional pythagorean aperiodic lattices. Optics Letters. 2021;**38**:1276

[16] Qidong F, Wang P, Huang C, Kartashov YV, Torner L, Konotop VV, et al. Optical soliton formation controlled by angle twisting in photonic moiré lattices. Nature Photonics. 2020; **14**(11):663-668

[17] Huang C, Ye F, Chen X, Kartashov YV, Konotop VV, Torner L. Localization-delocalization wavepacket transition in pythagorean aperiodic potentials. Scientific Reports. 2016;**6**(1): 32546

[18] Wang P, Zheng Y, Chen X, Huang C, Kartashov YV, Torner L, et al. Localization and delocalization of light

in photonic moiré lattices. Nature. 2020; 577(7788):42-46

[19] Bağcı M, Bakırtaş İ, Antar N. Fundamental solitons in parity-time symmetric lattice with a vacancy defect. Optical Communication. 2015;356: 472-481

[20] Bağcı M. Effects of lattice frequency on vacancy defect solitons in a medium with quadratic nonlinear response. Bitlis Eren Üniversitesi Fen Bilimleri Dergisi. 2022;11:344-351

[21] Bağcı M, Bakırtaş İ, Antar N. Vortex and dipole solitons in lattices possessing defects and dislocations. Optical Communication. 2014;331:204-218

[22] Martin H, Eugenieva ED, Chen Z, Christodoulides DN. Discrete solitons and soliton-induced dislocations in partially coherent photonic lattices. Physical Review Letters. 2004;92: 123902

[23] Mark J, Biondini G, Blair S. Localized multi-dimensional optical pulses in non-resonant quadratic materials. Mathematical Computational Simulation. 2001;56(6):511-519

[24] Bağcı M, Bakırtaş İ, Antar N. Lattice solitons in nonlinear Schrödinger equation with coupling-to-a-mean-term. Optical Communication. 2017; 383:330-340

[25] Bağcı M, Kutz JN. Spatiotemporal mode locking in quadratic nonlinear media. Physical Review A. 2020;102: 022205

[26] Buryak AV, Di Trapani P, Skryabin DV, Trillo S. Optical solitons due to quadratic nonlinearities: from basic physics to futuristic applications. Physical Reports. 2002;370(2):63-235

[27] Hayata K, Koshiba M. Multidimensional solitons in quadratic nonlinear media. Physical Review

Letters. 1993;71(20):3275-3278. DOI: 10.1103/PhysRevLett.71.3275

[28] Torner L, Sukhorukov AP. Quadratic solitons. Optical Photonic News. 2002;13(2):42-47

[29] Torruellas WE, Wang Z, Hagan DJ, VanStryland EW, Stegeman GI, Torner L, et al. Observation of two-dimensional spatial solitary waves in a quadratic medium. Physical Review Letters. 1995;74:5036

[30] Crasovan L-C, Torres JP, Mihalache D, Torner L. Arresting wave collapse by wave self-rectification. Physical Review Letters. 2003;91: 063904

[31] Schiek R, Pertsch T. Absolute measurement of the quadratic nonlinear susceptibility of lithium niobate in waveguides. Optical Material Express. 2012;2(2):126-139

[32] Benney DJ, Roskes GJ. Wave instabilities. Studies in Application Mathematics. 1969;48:377-385

[33] Davey A, Stewartson K. On three-dimensional packets of surface waves. Proceedings of the Royal Society A. 1974;338:101-110

[34] Ablowitz MJ, Haberman R. Nonlinear evolution equations—two and three dimensions. Physical Review Letters. 1975;35:1185-1188

[35] Djordjevic VD, Redekopp LG. On two-dimensional packets of capillary-gravity waves. Journal of Fluid Mechanics. 1977;79(4):703-714. DOI: 10.1017/S0022112077000408

[36] Mark J, Biondini G, Blair S. Multi-dimensional pulse propagation in non-resonant materials. Physical Review Letters. 1997;236(5):520-524

[37] Mark J, Biondini G, Blair S. Nonlinear Schrödinger equations with

mean terms in nonresonant multidimensional quadratic materials. Physical Review. E. 2001;**63**:046605

[38] Michael L. Sundheimer. Cascaded second-order nonlinearities in waveguides [PHD thesis]. 1994

[39] Ablowitz MJ. Wave collapse in a class of nonlocal nonlinear Schrödinger equations. Physica D: Nonlinear Phenomena. 2005;**207**(3):230-253

[40] Kelley PL. Self-focusing of optical beams. Physical Review Letters. 1965;**15**: 1005-1008

[41] Papanicolaou G, McLaughlin D, Weinstein M. Focusing singularity for the nonlinear Schrödinger equation. In: Fujita H, Lax PD, Strang G, editors. Nonlinear Partial Differential Equations in Applied Science; Proceedings of The U.S.-Japan Seminar, Tokyo, 1982, volume 81 of North-Holland Mathematics Studies. North-Holland; 1983

[42] Merle FH, Raphael P. On universality of blow-up profile for l2 critical nonlinear schrödinger equation. Inventiones Mathematicae. 2004;**156**: 565

[43] Moll KD, Gaeta AL, Fibich G. Self-similar optical wave collapse: Observation of the townes profile. Physical Review Letters. 2003;**90**: 203902

[44] Bağcı M. Partially PT -symmetric lattice solitons in quadratic nonlinear media. Physical Review A. 2021;**103**: 023530

[45] Bağcı M. Vortex solitons on partially PT -symmetric azimuthal lattices in a medium with quadratic nonlinear response. Journal of Mathematical Sciences and Modelling. 2021;4:117-125

[46] Bağcı M, Horikis TP, Bakırtaş İ, Antar N. Lattice solitons in a nonlocal

nonlinear medium with self-focusing and self-defocusing quintic nonlinearity. In: Antar N, Bakırtaş İ, editors. The Nonlinear Schrödinger Equation. Rijeka: IntechOpen; 2022

[47] Abo-Shaeer JR, Raman C, Vogels JM, Ketterle W. Observation of vortex lattices in bose-einstein condensates. Science. 2001;**292**: 476-479

[48] Matthews MR, Anderson BP, Haljan PC, Hall DS, Wieman CE, Cornell EA. Vortices in a bose-einstein condensate. Physical Review Letters. 1999;**83**:2498

[49] Bartal G, Fleischer JW, Segev M, Manela O, Cohen O. Two-dimensional higher-band vortex lattice solitons. Optics Letters. 2004;**29**:2049

[50] Fleischer JW, Bartal G, Cohen O, Manela O, Segev M, Hudock J, et al. Observation of vortex-ring "discrete" solitons in 2d photonic lattices. Physical Review Letters. 2004;**92**:3

[51] Freedman B, Bartal G, Segev M, Lifshitz R, Christodoulides DN, Fleischer JW. Wave and defect dynamics in nonlinear photonic quasicrystals. Nature. 2006;**440**:7088

[52] Kartashov YV, Malomed BA, Torner L. Solitons in nonlinear lattices. Reviews of Modern Physics. 2011;**83**: 247-305

[53] Yuri S. Dark optical solitons: physics and applications. Physics Reports. 1998; **298**:81-197

[54] Leblond H, Malomed BA, Mihalache D. Spatiotemporal vortex solitons in hexagonal arrays of waveguides. Physical Review E. 2011;**83**: 063825

[55] Izdebskaya YV, Shvedov VG, Jung PS, Krolikowski W. Stable vortex

soliton in nonlocal media with orientational nonlinearity. Optics Letters. 2018;**43**:66

[56] Chen Y-F, Beckwitt K, Wise FW, Aitken BG, Sanghera JS, Aggarwal ID. Measurement of fifth- and seventh-order nonlinearities of glasses. Journal of Optical Society America B. 2006;**23**(2): 347-352

[57] Azzouzi F, Triki H, Grelu P. Dipole soliton solution for the homogeneous high-order nonlinear schrödinger equation with cubic-quintic-septic non-kerr terms. Applied Mathematical Modelling. 2015;**39**:3-1300

[58] Komarov A, Leblond H, Sanchez F. Quintic complex ginzburg-landau model for ring fiber lasers. Physical Review E. 2005;**72**:025604

[59] Alidou Mohamadou CG, Tiofack L, Kofané TC. Wave train generation of solitons in systems with higher-order nonlinearities. Physical Review E. 2010; **82**:016601

[60] Yang J, Lakoba TI. Universally-convergent squared-operator iteration methods for solitary waves in general nonlinear wave equations. Studies in Applied Mathematics. 2007;**118**(2): 153-197

[61] Yang J. Nonlinear Waves in Integrable and Nonintegrable Systems. Philadelphia: SIAM; 2010

[62] Vakhitov NG, Kolokolov AA. Stationary solutions of the wave equation in a medium with nonlinearity saturation. Radiophysics and Quantum Electronics. 1973;**16**(7): 783-789

[63] Harvey A, Weinstein MI. On the bound states of the nonlinear schrödinger equation with a linear potential. Physica D: Nonlinear Phenomena. 1988;**30**(1): 207-218

[64] Michael I. Weinstein, Modulational stability of ground states of nonlinear schrödinger equations. SIAM Journal on Mathematical Analysis. 1985;**16**(3): 472-491

Chapter 2

The Propagation of Vortex Beams in Random Mediums

Sekip Dalgac and Kholoud Elmabruk

Abstract

Vortex beams acquire increasing attention due to their unique properties. These beams have an annular spatial profile with a dark spot at the center, the so-called phase singularity. This singularity defines the helical phase structure which is related to the topological charge value. Topological charge value allows vortex beams to carry orbital angular momentum. The existence of orbital angular momentum offers a large capacity and high dimensional information processing which make vortex beams very attractive for free-space optical communications. Besides that, these beams are well capable of reducing turbulence-induced scintillation which leads to better system performance. This chapter introduces the research conducted up to date either theoretically or experimentally regarding vortex beam irradiance, scintillation, and other properties while propagating in turbulent mediums.

Keywords: vortex beams, random medium, turbulence, scintillation, optical communications

1. Introduction

Wave front dislocations, in other words, phase defects which consist of edge dislocations, screw dislocations and mixed edge-screw dislocations are firstly proposed by Nye and Berry as a new type of light field principle [1]. The screw dislocation most prevalently known as front dislocation which presents a phase singularity at the center of the beam with zero amplitude and indefinite phase. Also, when both the real and imaginary parts of the wave function (ψ) equal zero the phase singularity is observed. Due to the fact that light field possesses unique properties such as phase singularly or dislocations, it paves the way for modern optics which called singular optics. Optical vortices are the primary topic of the singular optics [2]. Allen in 1992 realized that a beam of photons can hold singularity with azimuthally phase structure $e^{il\Theta}$ and carry an orbital angular momentum (OAM) where l is the topological charge and Θ *is the* azimuth angle [3]. Vortex beams possess distinct optical properties compared to the other beam types since they carry OAM. These beams have introduced a great diversity in a wide range of applications namely optical manipulation, biomedical applications, micro-fabrication, imaging, and micro-mechanics. Furthermore, they played an important role in the new generation of optical communication where OAM is employed as a new modulation technique in the optical communication systems [4, 5]. Such a feature makes the beams carrying OAM a perfect solution for the increasing demand of larger bandwidth and higher data rates in a diversity of applications such as 5G and 6G communication links, laser satellite communications, and remote sensing. However, in these applications the propagation of the laser beam in a random medium, which represents the channel of the system, degrades the probability of error performance of the system [6–12].

Actually, the propagation of laser beams through a random medium is governed fundamentally by three main phenomena namely absorption, scattering and refractive-index fluctuations. While absorption, scattering, which are caused by constituent gas and particles in the medium, resulted in the energy dissipation [13, 14]. The refractive-index fluctuations named turbulence originate from the temperature differences and cause intensity fluctuations (scintillation) that degrade the probability of error performance of the wireless optical communication system. In case that turbulence presence, the beams involve in extra beam spreading, beam wander, and scintillation that greatly hamper the performance of the communication system. Consequently, understanding the effects of turbulent medium on the propagating beam is an important issue for the researchers that paves the way towards mitigating the limitations caused by turbulence [15, 16].

This chapter presents a detailed review of the conducted work up to date on the propagation of vortex beams through random mediums. Accordingly, Section 2 starts with the representation of different types of vortex beams. Then, followed by the theory of the propagation in a random medium in Section 3. Subsequently, Section 4 discusses the atmospheric turbulence effect on the fully and partially coherent vortex beams. In addition, it represents the scintillation properties of vortex beams. In Section 5, we evaluate coherent and partially coherent vortex beam properties in oceanic turbulence. Furthermore, it covers the scintillation effects on the vortex beams propagating oceanic turbulence medium. Finally, Section 6 sums up the chapter by concluding the advantages that vortex beams offer for optical communication systems through the degradation of turbulence effects.

2. Representations of vortex beams

In this part of the chapter, expressions of different vortex beams are given at the source plane on the fundamental coordinate systems, neither Cartesian (s_x, s_y) or radial(s, φ) [17, 18]. Firstly, the source field expression of Gaussian vortex beam is;

$$E(s,\varphi) = \left(\frac{s}{\alpha_s}\right)^l \exp\left(-\frac{s^2}{\alpha_s^2}\right) \exp\left(jl\varphi\right) \tag{1}$$

where α_s, and l represent the source size and topological charge respectively.

Besides that, the source field expression of elliptical Gaussian vortex beam [19] can be written with the Cartesian coordinate as follows;

$$E(s_x,s_y) = \left(\frac{s_x + je_s s_y}{\alpha_s}\right)^l \exp\left(-\frac{s_x^2 + \varepsilon_s^2 s_y^2}{\alpha_s}\right) \exp\left[jl\tan^{-1}\left(\frac{e_s s_y}{s_x}\right)\right] \tag{2}$$

ε_s is the degree of ellipticity. Another widely investigated beam type is the Laguerre Gaussian vortex beam which can be expressed as [20, 21];

$$E(s,\varphi) = \left(\frac{s}{\alpha_s}\right)^l \exp\left(-\frac{s^2}{\alpha_s^2}\right) L_n^m\left(\frac{s^2}{\alpha_s^2}\right) \exp\left(jl\varphi\right) \tag{3}$$

$L_n{}^m$ is the Laguerre polynomial, with a polynomial degree of n. If n > 0, Bessel function $J_n{}^m$ with orders can also generate vortex beams [22]. Thus, Bessel–Gaussian vortex beam can be written as;

$$E(s, \varphi) = exp\left(-\frac{s^2}{\alpha_s{}^2}\right) J_m\left(\frac{s}{\alpha_s}\right) exp\left(jl\varphi\right) \qquad (4)$$

J_m is the Bessel function order with m. Finally, Flat topped Gaussian vortex beam expressed with the related source field expression [23] as given;

$$E(s_x, s_y) = \frac{1}{N}\left(\frac{s_x + js_y}{\alpha_s}\right)^m \sum_{n=1}^{N} (-1)^{n-1}(Nn)\, exp\left(-n\frac{s_x{}^2 + s_y{}^2}{\alpha_s{}^2}\right) \qquad (5)$$

N indicates the order of flat-topped Gaussian vortex beam. Moreover, Hermite–Gaussian vortex beam can be written as the superposition of two orthogonally polarized components under paraxial approximation [24]. The equation of Hermite–Gaussian vortex beam can be written as;

$$E(s_x, s_y) = exp\left(-\frac{s_x{}^2 + s_y{}^2}{\alpha_s{}^2}\right)\left[H_{nx}\left(\frac{s_x}{\alpha_s}\right)H_{ny}\left(\frac{s_y}{\alpha_s}\right)\vec{s_x} + H_{mx}\left(\frac{s_x}{\alpha_s}\right)H_{my}\left(\frac{s_y}{\alpha_s}\right)\vec{s_y}\right]$$
$$(6)$$

where the orders of the Hermite polynomials such as nx, ny, mx, my in $H_{nx}(), H_{ny}(), H_{mx}(), H_{my}()$ can be introduced as odd integers to create the desired zero on-axis field amplitude, this way such combinations can be regarded as vortex beams.

The optical field of the sinh-Gaussian vortex beam in the source plane can be specified as given in [25];

$$E(\vec{s}, 0) = sinh\left[\Omega(x_0 + y_0)\right] exp\left(-\frac{x_0{}^2 + y_0{}^2}{w_0{}^2}\right)\left[x_0 + j\ sgn\ (l)y_0\right]^{|l|} \qquad (7)$$

$\vec{s}$ is the position vector, Ω denote the constant parameter of the hyperbolic sinusoidal part, where sgn (l) can be introduced as a symbolic function.

In addition to coherent vortex beams, there exist various important types of partially coherent vortex beams in the literature. The cross spectral density (CSD) of partially coherent beams in the source plane can be expressed by the following general form [17];

$$W(s_1, s_2) = \langle E_{(s_1)}E^*{}_{(s_2)}\rangle = A_{(s_1)}A_{(s_2)}\, exp\left[jl(\varphi_1 - \varphi_2)\right]g(s_1 - s_2) \qquad (8)$$

E_s and A_s are electric field and its amplitude, respectively. The angular bracket and the asterisk denote ensemble average and complex conjugate. s_1 and s_2 are the two arbitrary points in the source plane, $g(s_1 - s_2)$ is the correlation function between two arbitrary points. As the amplitude (A_s) of the beam changes, different kind of partially coherent beams are obtainable. The correlation function for the Gaussian distribution is [26];

$$g(s_1 - s_2) = exp\left[-\frac{(s_1 - s_2)}{2\delta_0{}^2}\right] \qquad (9)$$

That δ_0 indicates initial coherence width. In case of $\delta_0 \rightarrow \infty$, Eq. (9) tends to a fully coherent vortex beam, However, when $\delta_0 \rightarrow 0$, Eq. (9) reduces to an incoherent vortex beam. On the other hand the CSD function of Gaussian Schell-model (GSM) vortex beam in the source plane is written as [27];

$$E(s_1, s_2, \varphi_1, \varphi_2) = exp\left[-\frac{s_1^2 + s_2^2}{4\sigma_0^2} - \frac{s_1^2 + s^2 - 2s_1 s_2 \cos(\varphi_1 - \varphi_2)}{2\delta_0^2} + jl(\varphi_1 - \varphi_2)\right]$$

(10)

Where σ_0 indicates transverse beam width. Furthermore, the CSD of the partially coherent LG beam in the source plane is obtained as [28];

$$E(s_1, s_2, \varphi_1, \varphi_2) = \left(\frac{\sqrt{2}s_1}{\omega_0}\right)^m \left(\frac{\sqrt{2}s_2}{\omega_0}\right)^m L_n{}^m \left(\frac{2s_1^2}{\omega_0^2}\right) L_p{}^m \left(\frac{2s_2^2}{\omega_0^2}\right) exp - \left(\frac{s_1^2 + s_2^2}{4\omega_0^2}\right)$$

$$\times exp\left[-\frac{s_1^2 + s_2^2 - 2s_1 s_2 \cos(\varphi_1 - \varphi_2)}{2\delta_0^2}\right] exp\left[jl(\varphi_1 - \varphi_2)\right] \quad (11)$$

$L_n{}^m$ is Laguerre polynomial with mode orders n and m. In the case that n = 0, Eq. (11) becomes a partially coherent LG_{01} beam. However, having the both mode ordersn and m set to zero, the beam turns to the well-known GSM beam. Furthermore, Laguerre Gaussian correlated Schell-model vortex (LGCSMV) beam in the source plane as special kind of the correlated partially coherent vortex beams, can be expressed as [29].

$$E(s_1, s_2) = exp\left[-\frac{s_1^2 + s_2^2}{4\sigma_0^2} - \frac{(s_1 - s_2)^2}{2\delta_0^2}\right] L_n{}^0 \frac{(s_1 - s_2)^2}{2\delta_0^2} exp\left[jl(\theta_1 - \theta_2)\right] \quad (12)$$

3. Theoretical background of beam propagation through random medium

In this part of the chapter, atmospheric and oceanic turbulence phenomena that influence the optical laser beam propagation are explained. Also, the theoretical background regarding the laser beam propagation is provided.

3.1 Atmospheric turbulence

Atmosphere is a medium that surrounds the Earth which mainly consists of gaseous such as nitrogen, oxygen, water vapor, carbon dioxide, methane, nitrous oxide, and ozone. As the beam propagates through atmospheric medium, the change of atmosphere temperature and wind velocity results in variation of the atmosphere's refractive index. These changes simply called atmospheric turbulence. Atmospheric turbulence is a non-linear process that is governed by Navier–Stokes equations. Since solving such kind of equations is challenging, the statistical approaches are developed. One of the widely used approaches is Kolmogrov power spectrum model that is given below [30];

$$\Phi_n(k) = 0.033C_n^2\kappa^{-11/3}; \quad 1/L_0 \ll \kappa \ll 1/l_0 \quad (13)$$

C_n^2 indicates the refractive index structure and $\kappa = |K|$ is the scalar wave number. Kolmogrov power spectrum does ignore the effects of the inner (l_0) and outer scales (L_0) of the turbulence since outer scale is infinity and the inner scale is so small. However, more elaborated power spectrums are suggested by Tatarski and Von-Karman. Tatarski power spectrum is defined as [31];

$$\Phi_n(k) = 0.033 C_n^2 \kappa^{-11/3} \exp\left(-\frac{\kappa^2}{\kappa_m^2}\right); \quad \kappa \gg \frac{1}{L_0} \tag{14}$$

where $\kappa_m = 5.92/l_0$. If the limit $1/L_0 \to 0$ ($L_0 \to \infty$) then this spectrum has a singularity at $\kappa = 0$. If the inner and outer scales of the turbulence are considered, Von-Karman spectrum can be defined to model the turbulence as follows [32];

$$\Phi_n(\kappa) = 0.033 C_n^2 \kappa^{-11/3} \frac{\exp\left(-\kappa^2/\kappa_m^2\right)}{\left(\kappa^2 + \kappa_0^2\right)^{11/6}}; \quad 0 \leq \kappa < \infty \tag{15}$$

3.2 Oceanic turbulence

As it is stated above, optical turbulence refers to the index of refraction fluctuations, which is one of the most significant features of optical wave propagation. Depending on the medium type, external and internal effect, there are some distinctions among the index of refraction fluctuations. For instance, while temperature fluctuation is fundamental reason for atmospheric turbulence, refraction index variation in seawater is caused by not only temperature fluctuations but also fluctuations of salinity. For that reason, power spectrum of ocean that considers both temperature and salinity fluctuations was firstly proposed in 2000 [33]. Power spectrum of oceanic turbulence is given for homogeneous and isotropic underwater media as follows;

$$\Phi_n(\kappa) = 0.388$$
$$\times 10^{-8} \varepsilon^{-11/3} \left[+2.35(\kappa\eta)^{2/3}\right] \frac{\chi_T}{\varsigma^2} \left[\varsigma^2 \exp\left(-A_T\delta\right) + \exp\left(-A_S\delta\right) - 2\varsigma\exp(-A_{TS}\delta)\right] \tag{16}$$

ε is the rate of dissipation for turbulent kinetic energy per unit mass of fluid, χ_T is the rate of dissipation of mean square temperature. $A_T = 1.863 \times 10^{-2}$, $A_{TS} = 9.41 \times 10^{-3}\delta = 8.284(\kappa\eta)^{4/3}+12.987(\kappa\eta)^2$. ς is the relative strength of temperature and salinity fluctuations, and finally, η represents the Kolmogrov inner scale.

3.3 Turbulence Modeling

The behavior of optical beams propagating in random medium can be understood by characterizing the medium qualitatively and quantitatively. Huygens–Fresnel principle is one of the most important modeling types to characterize beam propagation in turbulent medium [34]. The average intensity distribution at the observation plane can be expressed via Huygens–Fresnel principle as Eq. (17);

$$\left\langle I\left(\vec{R}, L\right)\right\rangle = \frac{k^2}{(2\pi L)^2} \iint E_0\left(\vec{r}_1, 0\right) E_0{}^*\left(\vec{r}_2, 0\right)$$
$$\times \exp\left\{\frac{ik}{2L}\left[\left(R - \vec{r}_1\right)^2 - \left(R - \vec{r}_2\right)^2\right]\right\}$$
$$\times \left\langle \exp\left[\psi\left(R, \vec{r}_1\right) + \psi^*\left(R, \vec{r}_2\right)\right]\right\rangle d\vec{r}_1 d\vec{r}_2 \tag{17}$$

$\vec{R}$ denotes the position vector at the observation plane, $\vec{r}_1$ and $\vec{r}_2$ represent the position vectors at the source plane, k is the wave number, the asterisk denotes the complex conjugation, the < > indicates the ensemble average over the medium statistics covering the log-amplitude and phase fluctuations due to the turbulent atmosphere, $\psi\left(R, \vec{r}_1\right)$ is the random part complex phase of a spherical wave. Moreover, Rytov approximation is another type of turbulence modeling for weak atmospheric turbulence [35]. The beam at the receiver plane can be written in terms of Rytov approximation as given Eq. (18);

$$E(\rho, z) = -\frac{i}{\lambda z} \exp(ikz) \int_0^{2\pi} \int_0^{+\infty} E(r, \theta) \exp\left[\frac{ik}{2z}(\rho - r)^2\right] \exp[\psi(r, \rho, z)] r \, dr \, d\theta$$

(18)

$\psi(r, \rho, z)$ denotes the random part due to the turbulence and ρ is the position vector. Furthermore, Random Phase Screen method can be employed to modeling turbulence [36]. It is described by the spatial spectrum of phase fluctuations. The spectrum of random phase can be expressed as;

$$\Phi\left(q_x, q_y, q_z\right) = 2\pi k \Delta z \, \Phi\left(q_x, q_y\right)$$

(19)

4. Coherent vortex beams in turbulent media

Fluctuation of electric field between two or more points can be considered as coherence of the light beams. The effect of coherence parameters is analyzed by different research groups [37, 38]. The term of coherent vortex beam firstly was revealed by Coullet in 1989, then Allen found that vortex beams can carry OAM [39, 40]. Since the first exploration of the vortex beams, many studies utilized in numerous fields such as quantum information [41], optical processing [42], optical manipulation [43] and optical communication systems [44]. Thus, the propagation of fully coherent vortex beams in turbulent medium has been investigated intensively in the literature.

4.1 Atmospheric turbulence

Despite the great advantages of free-space optical communication (FSO) systems, the propagation of the laser beam in atmosphere limits the performance of these systems. Mitigating these effects can be achieved through understanding the behavior of the propagating beam under different atmosphere circumstances. Accordingly, the literature has significant investigations on this topic. Recently, vortex beams have become one of the beam types under concentration. Considering Laguerre Gaussian (LG) vortex beams, it is proved that, as the topological charge increases, LG beam undergoes less broadening as given **Figure 1a**. Also, it is obtained that LG vortex beam is less affected by the turbulence than Gaussian beam as a result of the numerical analysis in [17, 18]. Thus, Gaussian beam suffers from more broadening than LG vortex beam. Moreover, Algebraic sum of the topological charges of LG beam is determined. Accordingly, the phase singularities existing in test aperture is approximately equal to the topological charge of the input LG vortex beam [48]. Fiber coupling of LG vortex beam in turbulent atmosphere is investigated by a theoretical model. LG beam that have small OAM number, low radial

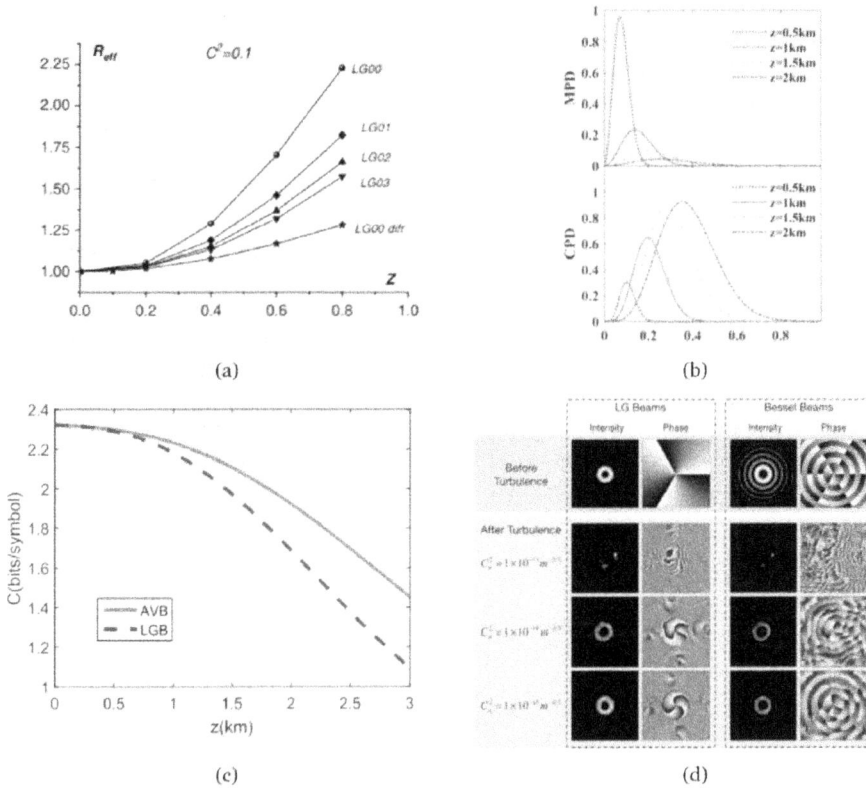

Figure 1.
Influence of (a) topological charge and (b) mode probability density (MPD) and crosstalk probability density of low-order LG beams for the various propagation distance for angular mode = 1. (c) the capacity of wireless optical links using AV beams versus LG beams, and (d) effect of turbulence on the intensity and phase distributions of Bessel vortex beams versus LG vortex beams [17, 45–47].

index and long wavelength gives higher coupling efficiency [49]. Mode probability density (MPD) of LG beam propagating in atmospheric turbulence is analyzed. MPD of LG vortex beam decreases while the distance increases as given **Figure 1b**. Additionally, MPD is increases by lower radial and waist radius, lower refractive index constant and shorter propagation distance [45]. The propagation properties of synthesized vortex beams compared with LG beams in free-space and in atmosphere is explored numerically. Propagation properties of LG beam shows the same characteristics with those of the synthesized vortex beams [50]. Furthermore, spiral spectrum of LG vortex beam and Anomalous vortex beam (AVB) is studied in details. It is achieved that; effects of atmospheric turbulence on LG vortex beam are more than those on Anomalous vortex beams as illustrated **Figure 1c**. Also, the spiral spectrum of the AVB is less affected by the turbulent atmosphere compared with LG vortex beam, in the case that AVB has larger beam order, longer wavelength, smaller topological charge, and at smaller refractive index structure constant, also propagating shorter distances [46]. Different kinds of vortex beams, including LG vortex beam and Bessel vortex beam were analyzed under the same turbulence conditions as given **Figure 1d**. Bessel vortex beams are more affected by the turbulence than LG vortex beams under the same circumstances [47]. It was experimentally demonstrated that, LG vortex beam exhibit enhanced backscatter (EBS) when only having even topological charge and LG beam may convert into corresponding Hermite Gaussian (HG) mode [51].

In addition researchers have analyzed Bessel Vortex Beams (BVB) in atmospheric turbulence. The degree of coherence of Bessel vortex beam decreases much faster under higher levels of fluctuation in the atmosphere [52]. The mean intensity of BVB versus dimensionless parameter (ξ) is given **Figure 2a**. It is obvious from the figure that increasing the topological charge results in decreasing the mean intensity of BVB [22]. Also, Bessel-Gaussian Vortex beams (BGV) have been studied numerically and experimentally, where it is observed that the OAM mean value does not show any variation during the propagation in atmospheric turbulence [56]. The mean intensity of BGV beams possessing phase singularities versus wavelength is given **Figure 2b**. It is clear that the central hole and the dark ring of the beams are gradually filled with the decrease of wavelength. Also, mean intensity of BGV beam decreases faster as the beam operating at a shorter wavelength or having either a narrower beam width, or a smaller topological charge [53]. Finally, a comparison between LG beam and BGV beam is conducted in terms of transmission quality and stability. According to the study given in [57], transmission quality and stability of BGV beam were observed to be better than those of the LG beams. Gaussian Vortex (GV) beam is another beam type that investigated frequently by the researchers.

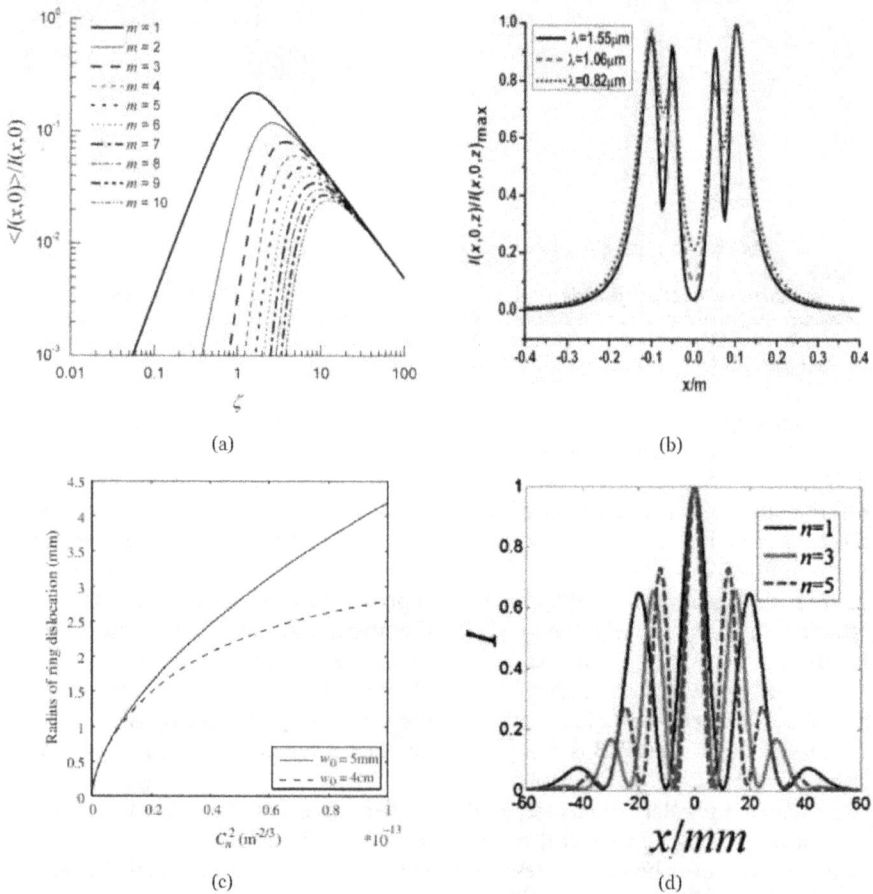

(a)

(b)

(c)

(d)

Figure 2.
(a) The average intensity of vortex Bessel beam versus dimensionless parameter and (b) BesselGaussian vortex beam with different wavelength, (c) illustration of the radius of a ring dislocation of vortex beam as a function of structure constant, and (d) beam order effect on the intensity distributions for four petal GV beam [22, 53–55].

GV beam enables us to calculate atmospheric turbulence strength by measuring radius of ring dislocations with different beam width as given **Figure 2c** [54]. Four Petal GV beam possessing high beam order undergoes transformation into more petals in the far field as achieved in **Figure 2d** [55].

The laser wavelength effect on the annular vortex beam is investigated when propagating in atmospheric turbulence [58]. It is observed that, operating at higher wavelengths causes lowering the central relative intensity and the central dark hollow is more achievable as stated in **Figure 3a**. Furthermore, beam width of a collimated vortex beam increases with the decrease of the wavelength [61]. Elliptically polarized (EP) vortex beams in turbulent atmosphere evolve into a Gaussian beam shape when the propagation distance is long enough and also flat-topped profile is obtained at a longer propagation distance as the topological charge increases [62]. Initial dark hollow profile of flat-topped vortex hollow beams remains the same in the short propagation distance then the beam evolves into a Gaussian-like beam under the strong turbulence [44]. Rectangular vortex beam array with arbitrary topological charge through atmospheric turbulence is analyzed and the obtained results clarify that beam array transform into a fan structure under moderate turbulence after propagating 1000 m, then turns to a single vortex beam after propagating 5000 m as given in **Figure 3b** and **c**. [59]. Also, optical vortex beams with higher topological charge are able to propagate longer distances in weak turbulent atmosphere. However, when the particular distance exceeds 500 km the output beam finally loses the vortex property and gradually becomes a Gaussian-shaped beam as illustrated in **Figure 3d** [60].

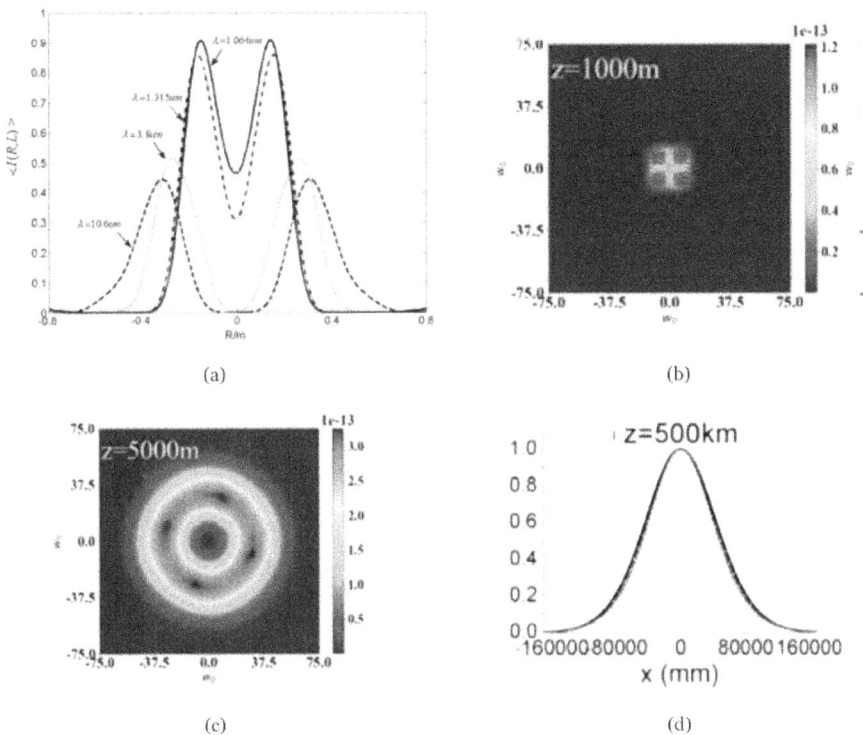

Figure 3.
(a) Average intensity of annular vortex beam with different beam wavelengths and rectangular vortex beam when distance (b) 1000 m and (c) 5000 m and optical vortex beam with distance of 500 km [58–60].

Furthermore, the influence of topological charge, wavelength, zenith, receiver aperture, waist radius, radial index and inner scale on spiral spectrum is investigated on the LG vortex beam propagating in slant atmospheric medium. It is achieved that, when propagation distance, topological charge, zenith and receive aperture increases, the spiral spectrum becomes wider. However, with the increase of wavelength and turbulence inner scale, the spiral spectrum spread less [63].

4.2 Scintillation properties

Optical wave propagating through a random medium such as the atmosphere, ocean and tissue etc. encounters fluctuations of beam intensity during the short and long propagation paths. This mechanism briefly explained by the scintillation of medium. Scintillation is caused by the external effect which is temperature variations in the random medium, resulting in index-of-refraction fluctuations (i.e., optical turbulence). Theoretical and experimental studies of scintillation have become more important nowadays since optical communication system adopts many types of beams. Accordingly, the scintillation index of LG beams is investigated in [17, 64]. The scintillation index of LG beam having different topological charges is demonstrated in **Figure 4a**. It is shown that, as propagation distance increases scintillation index increases as well. Also, it is obvious that the scintillation

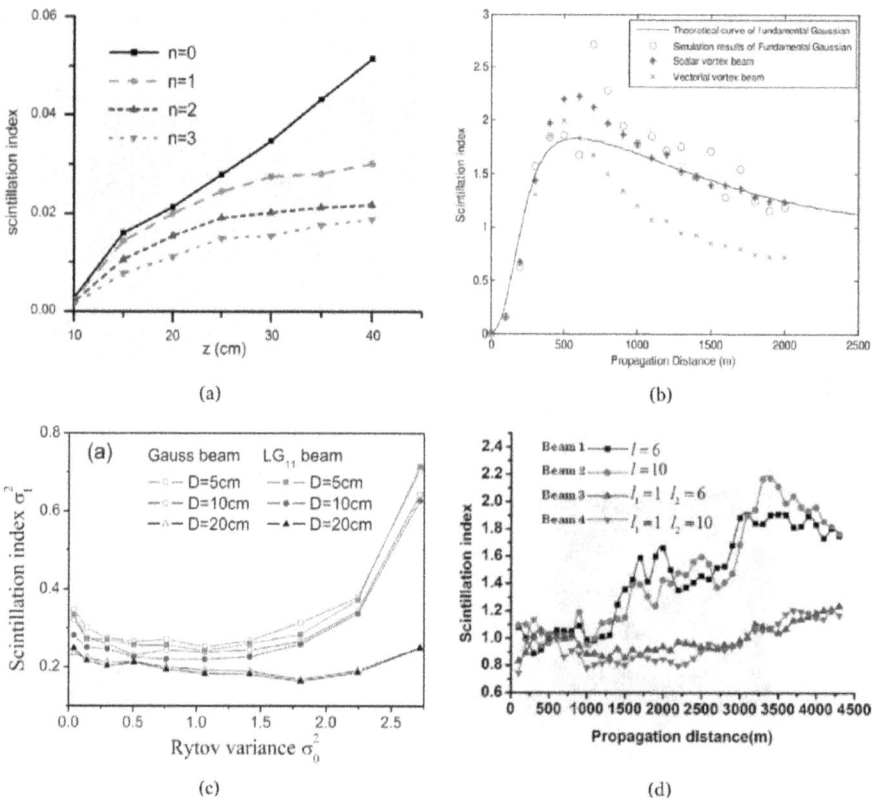

Figure 4.
Scintillation index of (a) LG beam with different beam orders (b) vector versus scalar vortex beams (c) LG vortex against Gaussian beam, (d) single and double vortex beam [21, 64–66].

of non-vortex beams is higher than that of the vortex beams since having a higher topological charge results in lower scintillation levels [64]. Also, it is obtained that, Gaussian beams are much more affected by the scintillation than LG vortex beams [17]. Additionally, the scintillation properties of vectorel and scalar vortex beam are analyzed both numerically and experimentally as shown in **Figure 4b**. This study realized that vectorel vortex beam provides an advantage over the scalar vortex beam since it has lower scintillation index for long propagation distances [21]. Furthermore, scintillation performance of various vortex beams (flat-topped Gaussian vortex, elliptical Gaussian vortex beam, Gaussian vortex beam) in strong turbulence region is investigated in [67]. It is achieved that, higher topological charges uniformly leading to lower scintillation [67]. The scintillation performance of Sinh Gaussian (SH-G) vortex beam has derived and investigated in [68]. This study has discovered that scintillation index of SH-G beam is higher than that of SH-G vortex beam under the same propagation circumstances. Comparison between Gauss and LG vortex beam in terms of scintillation index with different radius of targets is given in **Figure 4c**. The scintillation indices of the two beams decrease while weak turbulence effect exists. However, in case of strong turbulence, the scintillation indices increase. Moreover, the scintillation indices of Gaussian beam are higher than those of LG vortex beams [65]. Finally, **Figure 4d** shows the scintillation indices of single (beam 1 and beam 2) and double vortex beam (beam 3 and beam 4). All the beams have a similar scintillation levels at short propagation distance. On the other hand, the scintillation indices of the single vortex beams increase gradually at longer propagation distance [66]. Flat-topped Gaussian vortex beam propagating in a weakly turbulent atmosphere is investigated and scintillation properties are observed. It is found that flat-topped Gaussian vortex beam with high topological charges has less scintillation than the fundamental Gaussian beam [69].

4.3 Oceanic turbulence

Underwater Optical communication has attracted much attention due to its ability to provide the required large capacity and high-speed communication. Accordingly, many scientific research and exploration regarding the underwater environment are on progress. Among these, studying the propagation of laser beams under the effect of oceanic parameters namely spatial correlation length (σ), dissipation rate of temperature (χ_t), kinetic energy per unit mass of fluid (ε), relative strength of temperature, salinity fluctuations (ζ) and wavelength (λ). In this context, the detection probability characteristics of a Hyper geometric-Gaussian (HyGG) vortex beam propagating in oceanic turbulence are analyzed with different wavelengths as in **Figure 5a**. Beams operating at higher wavelengths have higher detection probability [70]. Further, HyGG beam with smaller topological charge is more resistant to oceanic turbulence. Furthermore, detection probability of Hermite Gaussian vortex beam tends to increase by the increase of ε [74]. Scintillation index of Gaussian vortex beam in oceanic turbulence is investigated for different waist widths as given in [75]. Besides that, Flat-topped vortex hollow beam is analyzed, where it recognized that this beam keeps its original intensity pattern in short propagation distances. Yet, it evolves into Gaussian like beam in far-field. Also, flat-topped vortex beam transforms into a Gaussian beam with decreasing of σ, ζ and ε as well as increasing of χ_t [76]. As given in **Figure 5b**, the detection ratio of Airy vortex beam is higher than that of LG beam when topological charge is higher than 5. Otherwise, it is the opposite when the topological charge is less than or equal 4. Likewise, the interference of Airy vortex beam becomes stronger when χ_t, ζ and the propagation distance increases [71]. Stochastic

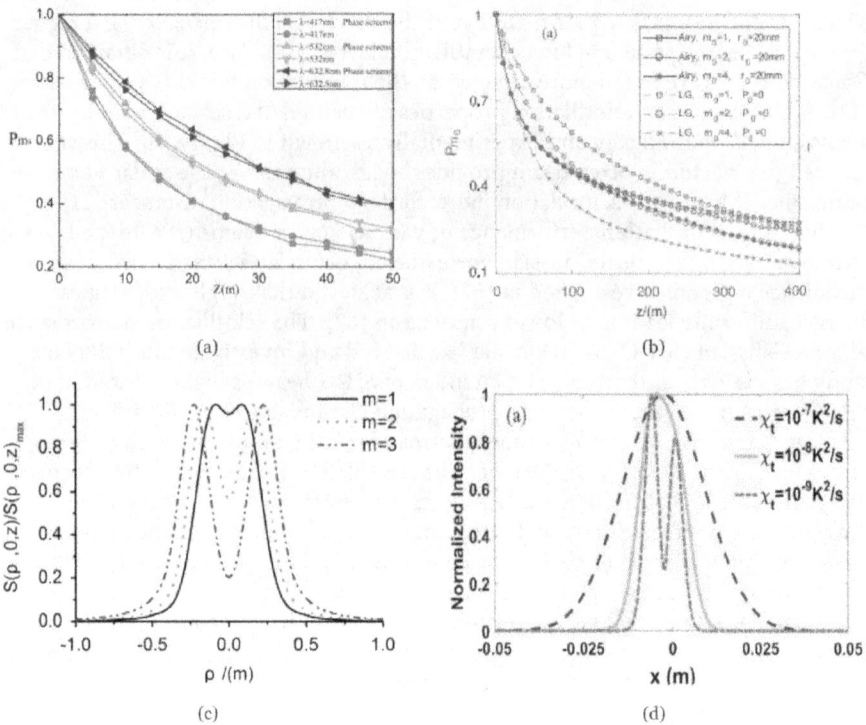

Figure 5.
(a) Detection probability of HyGG vortex beam with different wavelength and (b) detection ratio of LG versus airy vortex beam, (c) average intensity of stochastic electromagnetic vortex beam with different topological charges and (d) elliptical chirped Gaussian vortex with different value of χ_t [70–73].

electromagnetic vortex beam depending on the topological charge is analyzed in oceanic turbulence. While the topological charge increases, larger dark vortex core is obtained as in **Figure 5c** [72]. The effect of χ_t on rotating elliptical chirped Gaussian vortex beam (RECGVB) is illustrated in **Figure 5d**. It is obvious that, while χ_t increases, the minimum normalized intensity distribution of RECGVB increases and the spreading of the beam becomes wider which turn into Gaussian-like distribution at the receiver [73]. Finally, Lorentz-Gauss vortex beam propagating through oceanic turbulence is studied. As a result, it is obtained that beams with higher order topological charges have larger dark center and the beam can protect these properties as the distance increases [77].

4.4 Other mediums

Besides the oceanic and atmospheric medium, propagation of vortex beams in other mediums is important for the optical communication system. The propagation properties of Gaussian vortex beam in gradient index medium are investigated where the phase distribution of the beam is calculated by the Gradient index parameter. While the gradient index parameter increases, periodical cycles become shorter. The topological charge can also influence the period of the phase distributions [78]. Finally, anamalous vortex beam is investigated in strongly nonlocal nonlinear medium. The results present that, the input power plays a key role in the beam evolution. By selecting a proper input power, the beam width can be controlled [79].

5. Propagation properties of partially coherent vortex beams

A partially coherent beam is the beam with a low coherence length which was first demonstrated by Gori et al. [80]. This beam types have some unique properties, such as the cross-spectral density, and correlation function which is different than that of fully coherent beams. On the other hand, partially coherent beams are able to reduce the scintillation induced by the turbulence, the beam spreading, and the image noise when compared with the fully coherent beams [81, 82]. Recently, many research groups have conducted a wide range of studies regarding the propagation of partially coherent vortex beams either in atmosphere, ocean or other mediums.

5.1 Atmospheric turbulence

GSM vortex beam can be introduced as a partially coherent vortex (PCV) beam and many studies have inquired into this beam type. The influence of structure constant, spatial correlation length and beam index on GSM beam is investigated in details. As given in **Figure 6a** as the structure constant increases, the normalized propagation factor increases as well. Additionally, the beam width increases likewise [83, 86]. Similarly, multi GSM vortex beam with smaller correlation length tends to lose its dark hollow center and evolve into a Gaussian beam as obtained in **Figure 6b** [84, 87]. **Figure 6c** illustrates the scintillation index of GSM beam against

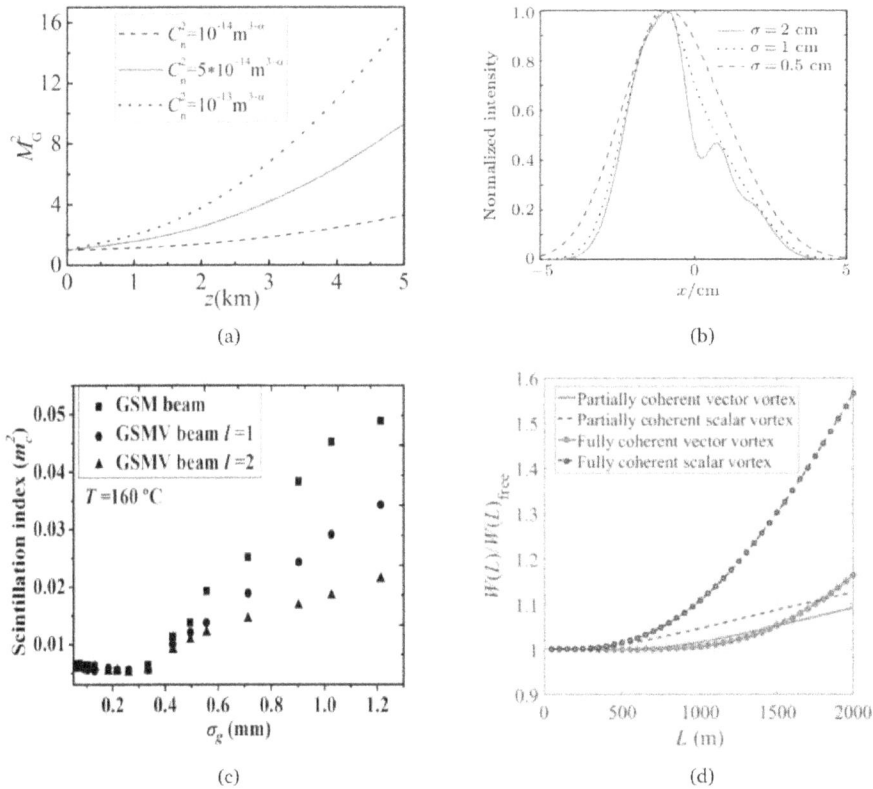

Figure 6.
(a) Average intensity of GSM vortex beam for different structure constant values, (b) average intensity of multi GSM vortex beam at different correlation lengths, (c) scintillation index of GSM beam against GSM vortex beams and (d) beam spreading of fully and partially coherent beam types [9, 83–85].

GSM vortex beam. It is clear that, the scintillation index of the two beams increases as the coherence length increases. Also, for the coherence length being larger than 0.35 mm, GSM vortex beam is less affected by the turbulence than the GSM beam [9]. Moreover, beam index is another important parameter that affect the GSM vortex beam. While the beam index increases, the focused beam profile becomes flatter [88, 89]. Finally, **Figure 6d** explains that, the PCV beam is obviously suffers from less beam spreading than the fully coherent vortex beam as expected [85].

Besides that, the propagation of partially coherent double-vortex beams in turbulent atmosphere is investigated deeply. Accordingly, it is observed that the topological charge, source beam width, degree of coherence at the source plane and the propagation distance are effective parameters on the intensity distributions. Consequently, as the propagation distance increases, beam profile changes to a Gaussian beam shape [90]. Moreover, the spreading of partially coherent flat-topped and Gaussian vortex beams in atmospheric turbulence is analyzed. It is achieved that the beam width of partially coherent beams increases as the distance increases and vortex beams are less affected by the atmospheric turbulence than the non-vortex ones. [91, 92]. Another study analyzes the partially four-petal elliptic Gaussian vortex beams propagating in turbulent atmosphere. It is achieved that partially coherent four-petal elliptic Gaussian beams with larger topological charge, smaller beam order, and larger ellipticity factor are less influenced by atmospheric turbulence. Moreover, vortex beams spread faster with the decreasing of the coherence length [93]. Scintillation index of partially coherent radially polarized vortex (PCRPV) beams, and PCV are analyzed as well. According to the obtained numerical results, scintillation index of PCRPV beams is lower than that of the PCV beams [94]. The propagation of partially coherent electromagnetic rotating elliptical Gaussian vortex (PCEREGV) beam through non-Kolmogorov turbulence is investigated numerically. Thus, it is realized that the normalized spectrum density of PCEREGV beam is slightly affected by the inner scale, while the operating wavelength greatly influences the spectrum density. Normalized spectrum density distributes more dispersedly and its minimum becomes larger when operating at higher wavelengths [95]. Finally, partially coherent twisted elliptical and circular vortex beams are analyzed and it is obtained that, elliptical vortex structure beam has advantage over the circular vortex with twisted phase modulation [96].

On the other hand, partially coherent LG and GSM vortex beams in slant atmospheric medium are analyzed by the researchers in [97, 98]. The beam wandering of GSM vortex beams along a slant path is lower than the horizontal path in case of long propagation distances [97]. Also, when partially coherent LG vortex beam is propagating in a slant path, bigger source coherence parameter causes a smaller transverse coherence length. A large zenith angle results in a small transverse coherence length of the beam [98].

5.2 Oceanic turbulence

Cross-spectral density and average intensity of GSM vortex beams propagating in oceanic turbulence are discussed and their analytical expressions are obtained using extended Huygens–Fresnel principle. The intensity equals zero at the center then as the distance increases, flat-topped beam takes place and, consequently, evolves into a Gaussian beam shape [99]. Not only the increase in χ_t, and ζ but also the decrease in ε lead the partially coherent GSM vortex beams to lose their dark hollow center pattern and evolve into a flat-topped beam and Gaussian-like beams as the propagation distance increases under the strong oceanic turbulence [87, 100]. In addition, Lorentz–Gauss vortex beam generated by a Schell-model source becomes wider with the increase of the oceanic turbulence parameters namely χ_t, and ς [101]. Furthermore, partially coherent flat-topped vortex hollow beam in oceanic turbulence with higher beam order loses its initial dark hollow center slower compared to the beam with lower beam order [23]. Partially coherent four-petal Gaussian vortex, anomalous hollow vortex beams are also discussed under the effect of oceanic turbulence. It is found that the partially coherent four-petal Gaussian vortex that has four petals profile in near field propagation, then turns into a Gauss-like beam rapidly with either decreasing σ, ς and ε, or increasing the oceanic parameter $χ_t$ in the far field [102]. For partially coherent anomalous hollow vortex beam, the parameters $χ_t$ and ς give rise to larger spreading of beam rather than ε [103].

5.3 Other mediums

The propagation of PCV beams in other mediums is also investigated. Consequently, the propagation properties of PCV beams in gain media are investigated. For longer propagation distances, PCV beams keep their original dark hollow intensity profile when having a higher topological charge value and larger coherence length. As the coherence length increases, the effective transmission distance of PCV beams with hollow distribution increases. However, fully coherent vortex beams always keep the hollow distribution while propagating in the gain medium [104].

6. Conclusion

The increasing importance of underwater and atmosphere wireless optical communication in a wide range of applications, has shaded the light on understanding the laser beam propagation in random media. In this context vortex beams play a role as one of the attractive laser beams which have become a widely investigated beam. The interest that these beams gained is due to their phase distribution that can be modulated to transmit the message signals. This way, they pose an alternative to the classical intensity or phase modulations that wireless optical communication links use. Thus, vortex beams are able to increase the ability of optical communication systems through mode multiplexing and high ratio terabit free-space data transmission. On the other hand, vortex beams are able to reduce the turbulence-induced scintillation, that leads to a better system performance.

In this context, this chapter introduces the research conducted up to date regarding the propagation of different vortex beam types in random medium. Besides summaries the effects of a variety of parameters such as the beam order, topological charge, coherence length, wavelength, source size, relative strength of temperature and salinity fluctuations on the beam properties. It observed that both Gaussian–Schell model vortex and elliptical vortex beams are able to improve the system performance through the reduction of scintillation that is induced by the atmospheric turbulence. Besides that, Laguerre–Gaussian vortex beam as an information carrier in the free-space optical link decreases the aperture averaged scintillation when increasing the topological charge value. The Laguerre–Gaussian vortex and combined Gaussian-vortex beams provides a room for the system performance improvement which is originated from the effective reduction of the scintillation index especially with the increase of the topological charge. Therefore, vortex beams are capable to propagate longer distances. In addition, beams with OAM mode provide another degree of freedom for multiplexing applications, especially space-division-multiplexing (SDM) systems which is sufficient for higher communication capacity. On the other hand, the double vortex beams offer advantages over the single vortex beams for long communication links. Moreover, a comparative study investigated the propagation of different of vortex beam types in strong turbulence, and revealed that as the values of topological charge increases the scintillation level decreases. Partially coherent vortex beams are able to reduce the scintillation, and beam spreading when compared to the fully coherent beams.

This chapter sets the models of optical wave propagating in random medium such as atmosphere, ocean and gain media. Then, focuses on the propagation of different vortex beams, either fully coherent or partially coherent, in different turbulent mediums. The presented results serve as an adequate database for understanding the propagation of vortex beams in random medium. Thus, provides an essential aid for further investigations in utilizing vortex beams in a wide range of application namely not only underwater optical communication, laser satellite communication systems but also sensing systems.

Author details

Sekip Dalgac and Kholoud Elmabruk*
Electrical Electronics Engineering Department, Sivas University of Science and Technology, Sivas, Turkey

*Address all correspondence to: elmabruk@sivas.edu.tr

IntechOpen

References

[1] Nye, John Frederick, Michael Victor Berry. Dislocations in wave trains. Proceedings of the Royal Society of London. A. Mathematical and Physical Sciences. 1974; 336.1605: 165-190.

[2] Soskin, M. S., M. V. Vasnetsov. Singular optics. *Progress in optics.* 2001; 42.4:219- 276.

[3] Allen, Les, et al. Orbital angular momentum of light and the transformation of Laguerre-Gaussian laser modes. *Physical review A.* 1992; 45.11: 8185.

[4] Graham, G. et al. Free-space information transfer using light beams carrying orbital angular momentum. *Optics express,* 2004; 12.22: 5448-5456.

[5] Čelechovský, Radek, Zdeněk Bouchal. Optical implementation of the vortex information channel. *New Journal of Physics.* 2007; 9.9: 328.

[6] Gbur, G. Robert K. Tyson. Vortex beam propagation through atmospheric turbulence and topological charge conservation. *JOSA A.* 2008; 25.1: 225-230.

[7] J. Wang et al., Terabit free-space data transmission employing orbital angular momentum multiplexing, Nature Photon. 2012; vol. 6, no. 7, Art. no. 488.

[8] Wang, Jian, et al. Terabit free-space data transmission employing orbital angular momentum multiplexing. *Nature photonics.* 2012; 6.7: 488-496.

[9] Liu, Xianlong, et al. Experimental demonstration of vortex phase-induced reduction in scintillation of a partially coherent beam. *Optics letters.* 2013; 38.24: 5323-5326.

[10] Sharma, Abhishek, et al. Analysis of 2× 10 Gbps MDM enabled inter satellite optical wireless communication under

the impact of pointing errors. *Optik.* 2021; 227: 165250.

[11] Cheng, Qiman, et al. An integrated optical beamforming network for two-dimensional phased array radar. *Optics Communications.* 2021; 126809.

[12] Emilien, Alvarez-Vanhard, Corpetti Thomas, and Houet Thomas. UAV & satellite synergies for optical remote sensing applications: A literature review. *Science of Remote Sensing.* 2021; 100019.

[13] Hohmann, Martin, et al. Direct measurement of the scattering coefficient. *Biomedical Optics Express.* 2021; 12.1: 320-335.

[14] Liu, Chang, and Lijun Xu. Laser absorption spectroscopy for combustion diagnosis in reactive flows: A review. *Applied Spectroscopy Reviews.* 2019; 54.1: 1-44.

[15] Gökçe, Muhsin Caner, and Yahya Baykal. Aperture averaging in strong oceanic turbulence. *Optics Communications.* 2018; 413:196-199.

[16] Li, Shuhui, et al. Atmospheric turbulence compensation in orbital angular momentum communications: Advances and perspectives. *Optics Communications.* 2018; 408:68-81.

[17] Lukin, V. P., Konyaev, P. A., & Sennikov, V. A. Beam spreading of vortex beams propagating in turbulent atmosphere. Applied optics, 2012; 51 (10), C84-C87.

[18] Wang, T., Pu, J., & Chen, Z. Beam-spreading and topological charge of vortex beams propagating in a turbulent atmosphere. *Optics Communications.* 2009; 282(7), 1255-1259.

[19] Liu, X., & Pu, J. Investigation on the scintillation reduction of elliptical

vortex beams propagating in atmospheric turbulence. *Optics express*. 2011; *19*(27), 26444-26450.

[20] Cheng, W., Haus, J. W., & Zhan, Q. (). Propagation of scalar and vector vortex beams through turbulent atmosphere. In *Atmospheric Propagation of Electromagnetic Waves III*. 2009; (Vol. 7200, p. 720004). International Society for Optics and Photonics.

[21] Cheng, W., Haus, J. W., & Zhan, Q. Propagation of vector vortex beams through a turbulent atmosphere. *Optics express*. 2009; *17* (20), 17829-17836.

[22] Lukin, I. P. Formation of a ring dislocation of a coherence of a vortex optical beam in turbulent atmosphere. In *Eleventh International Conference on Correlation Optics*. 2013; (Vol. 9066, p. 90660Q). International Society for Optics and Photonics.

[23] Liu, D., Wang, Y., & Yin, H. Evolution properties of partially coherent flat-topped vortex hollow beam in oceanic turbulence. *Applied optics*. 2015; *54*(35), 10510-10516.

[24] Eyyuboğlu, H. T. Hermite-cosine-Gaussian laser beam and its propagation characteristics in turbulent atmosphere. *JOSA A*. 2005; *22*(8), 1527-1535.

[25] Li, J., Zeng, J., Duan, M. Classification of coherent vortices creation and distance of topological charge conservation in non-Kolmogorov atmospheric turbulence. *Optics Express*. 2015; *23*(9), 11556-11565.

[26] Wolf, E. Introduction to the Theory of Coherence and Polarization of Light. 2007; Cambridge University Press.

[27] Wang, F., Zhu, S., Cai, Y. Experimental study of the focusing properties of a Gaussian Schell-model vortex beam. *Optics letters*. 2011; *36*(16), 3281-3283.

[28] Yang, Y., Chen, M., Mazilu, M., Mourka, A., Liu, Y. D., & Dholakia, K. Effect of the radial and azimuthal mode indices of a partially coherent vortex field upon a spatial correlation singularity. *New Journal of Physics*. 2013; *15*(11), 113053.

[29] Chen, Y., Wang, F., Zhao, C., & Cai, Y. Experimental demonstration of a Laguerre-Gaussian correlated Schell-model vortex beam. *Optics express*. 2014; *22*(5), 5826-5838.

[30] Du, W., Cheng, X., Wang, Y., Jin, Z., Liu, D., Feng, S., & Yang, Z. Scintillation Index of a Plane Wave Propagating Through Kolmogorov and Non-Kolmogorov Turbulence along Laser-Satellite Communication Downlink at Large Zenith Angles. *Journal of Russian Laser Research*. 2020, *41*(6), 616-627.

[31] Clifford, S. F. Temporal-frequency spectra for a spherical wave propagating through atmospheric turbulence. *JOSA*. 1971; *61*(10), 1285-1292.

[32] Von Karman, T. Progress in the statistical theory of turbulence. *Proceedings of the National Academy of Sciences of the United States of America*. 1948; *34*(11), 530.

[33] Nikishov, V. V., & Nikishov, V. I. Spectrum of turbulent fluctuations of the sea-water refraction index. *International journal of fluid mechanics research*. 2000; *27*(1).

[34] Chu, X., & Zhou, G. Power coupling of a two-Cassegrain-telescopes system in turbulent atmosphere in a slant path. *Optics express*. 2007; *15*(12), 7697-7707.

[35] Potvin, G. General Rytov approximation. *JOSA A*. 2015; *32*(10), 1848-1856.

[36] Bayraktar, M. Properties of hyperbolic sinusoidal Gaussian beam propagating through strong atmospheric

turbulence. *Microwave and Optical Technology Letters*. 2021; *63*(5), 1595-1600.

[37] Zhou, X., Pang, Z., & Zhao, D. Partially coherent Pearcey–Gauss beams. *Optics Letters*. 2020; *45*(19), 5496-5499.

[38] Tang, S., Yan, J., Yong, K., & Zhang, R. Propagation characteristics of vortex beams in anisotropic atmospheric turbulence. *JOSA B*. 2020; *37*(1), 133-137.

[39] Coullet, P., Gil, L., & Rocca, F. Optical vortices. *Optics Communications*. 1989; *73*(5), 403-408.

[40] Allen, L., &Beijersbergen, M. W. RJC Spreeuw, and JP Woerdman. *Phys. Rev. A*. 1992; *45*, 8185.

[41] Perez-Garcia, B., Francis, J., McLaren, M., Hernandez-Aranda, R. I., Forbes, A., & Konrad, T. Quantum computation with classical light: The Deutsch Algorithm. *Physics Letters A*. 2015; *379* (28-29), 1675-1680.

[42] Wang, J., Yang, J. Y., Fazal, I. M., Ahmed, N., Yan, Y., Huang, H., Willner, A. E. (2012). Nat. Photonics. 2012; 6 (7), 488.

[43] He, H., Friese, M. E. J., Heckenberg, N. R., &Rubinsztein-Dunlop, H. Direct observation of transfer of angular momentum to absorptive particles from a laser beam with a phase singularity. *Physical review letters*. 1995; *75* (5), 826.

[44] Wang, J. Advances in communications using optical vortices. *Photonics Research*, *4*(5), B14-B28. strong perturbation media. *Communications in Computational Physics*. 2016; *submitted*.

[45] Li, Y., Han, Y., Cui, Z., & Hui, Y. Probability density performance of Laguerre-Gaussian beams propagating in non-Kolmogorov atmospheric turbulence. *Optik*. 2018; *157*, 170-179.

[46] Li, F., Lui, H., &Ou, J. Spiral spectrum of anomalous vortex beams propagating in a weakly turbulent atmosphere. *Journal of Modern Optics*. 2020; *67*(6), 501-506.

[47] Fu, S., & Gao, C. Influences of atmospheric turbulence effects on the orbital angular momentum spectra of vortex beams. *Photonics Research*. 2016; *4*(5), B1-B4.

[48] Ge, X. L., Wang, B. Y., & Guo, C. S. Evolution of phase singularities of vortex beams propagating in atmospheric turbulence. *JOSA A*. 2015; *32*(5), 837-842.

[49] Yu, L., & Zhang, Y. Investigation on the Coupling of Vortex Beam into Parabolic Fiber in Turbulent Atmosphere. *IEEE Photonics Journal*. 2019; *11*(6), 1-8.

[50] Aksenov, V. P., Dudorov, V. V. E., &Kolosov, V. V. Properties of vortex beams formed by an array of fibre lasers and their propagation in a turbulent atmosphere. *Quantum Electronics*. 2016; *46*(8), 726.

[51] Yu, J., Huang, Y., Gbur, G., Wang, F., & Cai, Y. Enhanced backscatter of vortex beams in double-pass optical links with atmospheric turbulence. *Journal of Quantitative Spectroscopy and Radiative Transfer*. 2019; *228*, 1-10.

[52] Lukin, I. P. Mean intensity of vortex Bessel beams propagating in turbulent atmosphere. *Applied optics*. 2014; *53* (15), 3287-3293.

[53] Yue, X., Ge, X., Lyu, Y., Zhao, R., Wang, B., Han, K., Fu, S. Mean intensity of lowest order Bessel-Gaussian beams with phase singularities in turbulent atmosphere. *Optik*. 2020; *219*, 165215.

[54] Gu, Y., & Gbur, G. Measurement of atmospheric turbulence strength by vortex beam. *Optics Communications*. 2010; *283*(7), 1209-1212.

[55] Guo, L., Tang, Z., & Wan, W. Propagation of a four-petal Gaussian vortex beam through a paraxial ABCD optical system. *Optik*. 2014; *125*(19), 5542-5545

[56] Lukin, I. P. Integral momenta of vortex Bessel–Gaussian beams in turbulent atmosphere. *Applied optics*. 2016; *55*(12), B61-B66.

[57] Li, Y. Q., Wang, L. G., & Wu, Z. S. (). Study on intensities, phases and orbital angular momentum of vortex beams in atmospheric turbulence using numerical simulation method. *Optik*. 2018; *158*, 1349-1360.

[58] Wu, H., Xiao, R., Li, X., Sun, Z., Wang, H., Xu, X., & Wang, Q. Annular vortex beams with apertures and their characteristics in the turbulent atmosphere. *Optik*. 2015; *126*(23), 3673-3677.

[59] Luo, C., Lu, F., & Han, X. E. Propagation and evolution of rectangular vortex beam array through atmospheric turbulence. *Optik*. 2020; *218*, 164913.

[60] Wang, L. G., & Zheng, W. W. The effect of atmospheric turbulence on the propagation properties of optical vortices formed by using coherent laser beam arrays. *Journal of Optics A: Pure and Applied Optics*. 2009; *11*(6), 065703.

[61] Luo, C., & Han, X. E. Evolution and Beam spreading of Arbitrary order vortex beam propagating in atmospheric turbulence. *Optics Communications*. 2020; *460*, 124888.

[62] Ou, J., Jiang, Y., & He, Y. Intensity and polarization properties of elliptically polarized vortex beams in turbulent atmosphere. *Optics & Laser Technology*. 2015; *67*, 1-7.

[63] Liu, Z., Wei, H., Cai, D., Jia, P., Zhang, R., & Li, Z. Spiral spectrum of Laguerre-Gaussian beams in slant non-

Kolmogorov atmospheric turbulence. *Optik*. 2017; *142*, 103-108.

[64] Chen, Z., Li, C., Ding, P., Pu, J., & Zhao, D. Experimental investigation on the scintillation index of vortex beams propagating in simulated atmospheric turbulence. *Applied Physics B*. 2012; *107*(2), 469-472.

[65] Li, Y., Wang, L., Gong, L., & Wang, Q. Speckle characteristics of vortex beams scattered from rough targets in turbulent atmosphere. *Journal of Quantitative Spectroscopy and Radiative Transfer*. 2020; *257*, 107342.

[66] Liu, Y., Zhang, K., Chen, Z., & Pu, J. Scintillation index of double vortex beams in turbulent atmosphere. *Optik*. 2019; *181*, 571-574.

[67] Eyyuboğlu, H. T. Scintillation behaviour of vortex beams in strong turbulence region. *Journal of Modern Optics*. 2016; *63*(21), 2374-2381.

[68] Zhang, Y., Zhou, X., & Yuan, X. Performance analysis of sinh-Gaussian vortex beams propagation in turbulent atmosphere. *Optics Communications*. 2019; *440*, 100-105.

[69] Elmabruk, K., Eyyuboglu, H. T. Analysis of flat-topped Gaussian vortex beam scintillation properties in atmospheric turbulence. *Optical Engineering*. 2019; *58*(6), 066115.

[70] Wang, X., Wang, L., Zhao, S. Research on Hypergeometric-Gaussian Vortex Beam Propagating under Oceanic Turbulence by Theoretical Derivation and Numerical Simulation. *Journal of Marine Science and Engineering*. 2021; *9*(4), 442.

[71] Wang, X., Yang, Z., & Zhao, S. Influence of oceanic turbulence on propagation of Airy vortex beam carrying orbital angular momentum. *Optik*. 2019; *176*, 49-55.

[72] Xu, J., Zhao, D. Propagation of a stochastic electromagnetic vortex beam in the oceanic turbulence. *Optics & laser technology*. 2014; *57*, 189-193.

[73] Ye, F., Zhang, J., Xie, J., & Deng, D. Propagation properties of the rotating elliptical chirped Gaussian vortex beam in the oceanic turbulence. *Optics Communications*. 2018; *426*, 456-462.

[74] Li, Y., Yu, L., & Zhang, Y. Influence of anisotropic turbulence on the orbital angular momentum modes of Hermite-Gaussian vortex beam in the ocean. *Optics express*. 2017; *25*(11), 12203-12215.

[75] Wang, H., Zhang, H., Ren, M., Yao, J., & Zhang, Y. Phase discontinuities induced scintillation enhancement: coherent vortex beams propagating through weak oceanic turbulence. *arXiv preprint arXiv*. 2021; *2102.03184*.

[76] Liu, D., Chen, L., Wang, Y., Wang, G., Yin, H. Average intensity properties of flat-topped vortex hollow beam propagating through oceanic turbulence. *Optik*. 2016; *127*(17), 6961-6969.

[77] Liu, D., Yin, H., Wang, G., & Wang, Y. Spreading of a Lorentz-Gauss Vortex Beam Propagating through Oceanic Turbulence. *Current Optics and Photonics*. 2019; *3*(2), 97-104

[78] Yang, S., Wang, J., Guo, M., Qin, Z., & Li, J. Propagation properties of Gaussian vortex beams through the gradient-index medium. *Optics Communications*. 2020; *465*, 125559.

[79] Dai, Z., Yang, Z., Zhang, S., & Pang, Z. Propagation of anomalous vortex beams in strongly nonlocal nonlinear media. *Optics Communications*. 2015; *350*, 19-27.

[80] Gori, F., Santarsiero, M., Borghi, R., Vicalvi, S. Partially coherent sources with helicoidal modes. *Journal of Modern Optics*. 1998; *45*(3), 539-554.

[81] Dogariu, A., Amarande, S. Propagation of partially coherent beams: turbulence-induced degradation. *Optics letters*. 2003; *28*(1), 10-12.

[82] Dong, M., Zhao, C., Cai, Y. et al. Partially coherent vortex beams: Fundamentals and applications. Sci. China Phys. Mech. Astron. 2021; 64, 224201.

[83] Li, J., Wang, W., Duan, M., Wei, J. Influence of non-Kolmogorov atmospheric turbulence on the beam quality of vortex beams. *Optics Express*. 2016; *24*(18), 20413-20423.

[84] Song, Y. S., Dong, K. Y., Chang, S., Dong, Y., & Zhang, L. Properties of off-axis hollow Gaussian-Schell model vortex beam propagating in turbulent atmosphere. *Chinese Physics B*. 2020; *29* (6), 064213.

[85] Cheng, M., Guo, L., Li, J., Li, J., & Yan, X. Enhanced vortex beams resistance to turbulence with polarization modulation. *Journal of Quantitative Spectroscopy and Radiative Transfer*. 2019; *227*, 219-225.

[86] Dong, Y., Dong, K., Chang, S., Song, Y. Propagation of rectangular multi-Gaussian Schell-model vortex beams in turbulent atmosphere. *Optik*. 2020; *207*, 163809.

[87] Ma, X., Wang, G., Zhong, H., Wang, Y., & Liu, D. The off-axis multi-Gaussian Schell-model hollow vortex beams propagation in free space and turbulent ocean. *Optik*. 2021; *228*, 166180.

[88] Tang, M., & Zhao, D. Propagation of multi-Gaussian Schell-model vortex beams in isotropic random media. *Optics express*. 2015; *23*(25), 32766-32776

[89] Zhang, Y., Liu, L., Zhao, C., & Cai, Y. Multi-Gaussian Schell-model vortex beam. *Physics Letters A*. 2014; *378*(9), 750-754.

[90] Fang, Guijuan, et al. Propagation of partially coherent double-vortex beams in turbulent atmosphere. *Optics & Laser Technology.* 2012; 44.6: 1780-1785.

[91] He, X., &Lü, B. Propagation of partially coherent flat-topped vortex beams through non-Kolmogorov atmospheric turbulence. *JOSA A.* 2011; 28(9), 1941-1948.

[92] Wang, T., Pu, J., & Chen, Z. Propagation of partially coherent vortex beams in a turbulent atmosphere. *Optical Engineering.* 2008; 47(3), 036002.

[93] Kenan Wu, Ying Huai, Tianliang Zhao, and Yuqi Jin. Propagation of partially coherent four-petal elliptic Gaussian vortex beams in atmospheric turbulence, Opt. Express. 2018; 26, 30061-30075.

[94] Yu, J., Huang, Y., Wang, F., Liu, X., Gbur, G., & Cai, Y. Scintillation properties of a partially coherent vector beam with vortex phase in turbulent atmosphere. *Optics express.* 2019; 27(19), 26676-26688.

[95] Zhang, J., Xie, J., & Deng, D. Second-order statistics of a partially coherent electromagnetic rotating elliptical Gaussian vortex beam through non-Kolmogorov turbulence. *Optics express.* 2018; 26(16), 21249-21257.

[96] Wang, L., Wang, J., Yuan, C., Zheng, G., Chen, Y. Beam wander of partially coherent twisted elliptical vortex beam in turbulence. *Optik.* 2020; 218, 165037.

[97] Li, J., Zhang, H., Lü, B. Partially coherent vortex beams propagating through slant atmospheric turbulence and coherence vortex evolution. *Optics & Laser Technology.* 2010; 42(2), 428-433.

[98] Lv, H., Ren, C., & Liu, X. Orbital angular momentum spectrum of partially coherent vortex beams in slant atmospheric turbulence. *Infrared Physics & Technology.* 2020; 105, 103181.

[99] Huang, Y., Zhang, B., Gao, Z., Zhao, G., Duan, Z. Evolution behavior of Gaussian Schell-model vortex beams propagating through oceanic turbulence. *Optics Express.* 2014; 22(15), 17723-17734.

[100] Liu, D., & Wang, Y. Properties of a random electromagnetic multi-Gaussian Schell-model vortex beam in oceanic turbulence. *Applied Physics B.* 2018; 124 (9), 1-9.

[101] Liu, D., Yin, H., Wang, G., & Wang, Y. Propagation of partially coherent Lorentz–Gauss vortex beam through oceanic turbulence. *Applied optics.* 2017; 56(31), 8785-8792.

[102] Liu, D., Wang, Y., Wang, G., Luo, X., & Yin, H. Propagation properties of partially coherent four-petal Gaussian vortex beams in oceanic turbulence. Laser Physics. 2016; 27(1), 016001.

[103] Liu, D., Wang, G., Yin, H., Zhong, H., & Wang, Y. Propagation properties of a partially coherent anomalous hollow vortex beam in underwater oceanic turbulence. *Optics Communications.* 2019; 437, 346-354.

[104] Guo, X., Yang, C., Duan, M., Guo, M., Wang, J., Li, J. Propagation of Partially Coherent Vortex Beams in Gain Media. *Optik.* 2021; 167361.

Section 2

Vortices in Fluids

Chapter 3

Searching of Individual Vortices in Experimental Data

Daniel Duda

Abstract

The turbulent flows consist of many interacting vortices of all scales, which all together self-organize being responsible for the statistical properties of turbulence. This chapter describes the searching of individual vortices in velocity fields obtained experimentally by Particle Image Velocimetry (PIV) method. The vortex model is directly fitted to the velocity field minimazing the energy of the residual. The zero-th step (which does not a priori use the vortex model) shows the velocity profile of vortices. In the cases dominated by a single vortex, the profile matches the classical solutions, while in turbulent flow field, the profile displays velocity decrease faster than $1/r$. The vortices fitted to measured velocity field past a grid are able to describe around 50 % of fluctuation energy by using 15 individual vortices, and by using 100 vortices, the fluctuating field is reconstructed by 75 %. The found vortices are studied statistically for different distances and velocities.

Keywords: vortex, turbulence, Particle Image Velocimetry, grid turbulence, individual vortex searching algorithm, vortex model

1. Introduction

Contemporary exploration of turbulent flows focuses on statistical characteristics [1] such as study of distributions [2, 3], Fourier analysis [4], correlations [5, 6], or the Proper Orthogonal Decompositions [7–9]. The success of statistical approach is declared by the large applicability of numerical simulations, which are able to perfectly match the experimental data. Although it is possible to predict the statistical development of turbulent flow, this is still far from *understanding* the turbulence. The turbulence consists of vortices [10] and other coherent structures [11] whose multi-body interactions are responsible for the life-like behavior—the flow can be *infected* by turbulence [12], and it dies when it is not fed [13]; turbulence fastens the energy transfer from low-entropy energy source to large-entropy energy (heat) by decreasing its own entropy via self-organization and the rise of coherent structures.

The importance of individual vortices to the turbulent statistics is best shown by the problem of *quantum turbulence* [14, 15], which consists of *quantized vortices* [16], which fulfill the Helmholtz circulation theorems [17]—their circulation is constant

IntechOpen

and equal to $\Gamma_{1qv} = \kappa = 2\pi\hbar/m_4 \approx 9.997 \cdot 10^{-8} \text{m}^2/\text{s}$, ($m_4$ is the mass of single helium 4 atom, it applies $2 \times m_3$ for helium 3 as it is a fermion); thus the vortices cannot end anywhere in the fluid, only at the fluid domain boundary, or they can form closed loops. The energy cascade can be realized only via vortex interactions, reconnections [16], and the helical Kelvin waves on the quantized vortices leading to phonon emission due to nonlinear interactions [18]. This nature of turbulence made of a tangle of identical vortices instead of different vortices as it is in Richardson cascade leads to polynomial velocity distribution [19] instead of almost-Gaussian distribution observed in classical turbulence [3, 20]. Despite this fact, the large-scale observation of superfluid flows shows the same picture as the classical flows do [21, 22]. The transition between both regimes depends on the length scale [2]. The interacting tangle of quantized vortices builds up the turbulence, whose structure is classical on large scale.

The fluid simulation by using the quantized vortices [23] is able to reconstruct the velocity spectra [18] and overall topology [24]. This method is applied in classical turbulence among others by the group of Ilia Marchevsky [25–27].

The behavior of individual vortices in experiment is studied by many groups; however, it is often limited to the case of some single vortex or vortex system dominating the flow. Among others, let us mention the work of Ben-Gida [28], who detected vortices in a wake past accelerating hydrofoil in stably stratified or mixed water. He used the maxima of λ_2 criterion [29]. De Gregorio [30] observed the tip vortex of helicopter rotor blade, and for its detection used the Γ_2 criterion [31]. They measured the vortex velocity profile and found that it is similar to the Vatistas model (discussed later here); they studied the development of the tip vortex and observed the interactions of tip vortices of various blades and various turn ages downstream the helicopter jet. Graftieaux et al. [31] developed the the functions Γ_1 and Γ_2 for the study of swirling flow in a duct. They detected a single vortex in each snapshot and in average field measuring the distribution of the distance of average and instantaneous vortex center. Their scalar function used for vortex detection is effectively similar to smoothed circulation over some neightborhood; therefore it nicely solves the issue of all experimental data: the noise; on the other hand, it introduces a new artificial parameter of the detection: the neighborhood area. Kolář [32] developed probably the most accurate criterion for identifying the three components of velocity gradient tensor—the shear, strain, and rotation. But his method needs a large number of transformations in each point. Maciel et al. [33] noticed that eigen axes of the velocity gradient tensor might do the same job.

In this chapter, the method is based on direct fitting of the instantaneous velocity field by some vortex model with scalar criterion used for the prefit only. In the next section, the available vortex models are introduced, then the velocity profiles on experimental data are shown introducing a new vortex model. Later the prefit function and the fitting procedure are shown, and at the end, some results obtained in the grid turbulence are presented.

1.1 Vortex profiles

A principal disadvantage of any fitting algorithm is the need of *a priori* knowledge of the functional dependence of the data, in our case, to know the vortex model fitted to the data. There have been a lot of different vortex models developed in the past. The basic idea of a vortex model is the circulation-free potential vortex, whose entire circulation is focused inside an infinitesimal topological singularity— the vortex filament. Everywhere else, the vorticity ω is zero.

$$u_\theta^{PV}(r) = \frac{\Gamma}{2\pi r} \tag{1}$$

The vorticity ω of this potential vortex is a scalar in the discussed simple two-dimensional case: $\omega = \left(\frac{\partial v}{\partial x} - \frac{\partial u}{\partial y}\right) = \frac{\Gamma}{2\pi}\left(\frac{1\cdot(x^2+y^2)-x\cdot(2x)}{(x^2+y^2)^2} - \frac{-1\cdot(x^2+y^2)+y\cdot(2y)}{(x^2+y^2)^2}\right) = \frac{\Gamma}{2\pi}\cdot\frac{1-1}{x^2+y^2}$

Among the infinite velocity of undefined direction in the center, the large velocity gradients smoothen the flow in a way, that there is minimum relative motion at small scales leading to the *solid-body rotation* with tangential velocity linearly increasing with the distance from the center

$$u_\theta^{SBR}(r) = \frac{\Gamma}{2\pi R}\cdot\frac{r}{R} \tag{2}$$

and vorticity ω being constant everywhere

$$\omega = \left(\frac{\partial v}{\partial x} - \frac{\partial u}{\partial y}\right) = \frac{\Gamma}{2\pi R^2}\left(\frac{\partial x}{\partial x} - \frac{\partial(-y)}{\partial y}\right) = \frac{\Gamma}{\pi R^2}$$

Simple connection of these two ideas is called Rankine vortex. A new parameter of the vortex is introduced: the vortex core radius R (in solid body rotation vortex (2), R played only the unit role: r/R is dimensionless distance, $\Gamma/2\pi R$ is tangential velocity at dimensionless distance $r/R = 1$). The fluid in this vortex rotates as a solid body inside the sharply bounded vortex core, while it orbits without internal rotation as a potential vortex outside of the circle bounded by R

$$u_\theta^{RV}(r) = \frac{\Gamma}{2\pi R}\cdot\begin{cases} r/R & \text{for } r < R \\ R/r & \text{for } r > R \end{cases} \tag{3}$$

Generally, there are not much sharp changes in the nature; therefore, a smooth solution is introduced by Oseen

$$u_\theta^{OV}(r) = \frac{\Gamma}{2\pi R}\cdot\frac{R}{r}\cdot\left(1 - e^{-(r/R)^2}\right) \tag{4}$$

This is one of the exact solutions of Navier-Stokes equations containing the temporal evolution as well, and it is called Lamb-Oseen vortex and then the core scales as $R \sim \sqrt{t}$ with time. However, we focus on descriptive analysis of instantaneous two-dimensional velocity fields observed experimentally without the temporal development. There exists more possible exact solutions of Navier-Stokes equations, let us mention at least the Burgers vortex, the Kerr-Dold vortex, or the Amromin vortex [34] with turbulent vortex core and potential envelope.

Mathematical simplification of Oseen vortex is suggested by Kaufmann [35] and later discovered independently by Scully et al. [36]. It uses just the first term of Taylor expansion of the exponential in the Oseen vortex, Eq. (4), as it is shown by Bhagwat and Leishman [37], and it is generalized by Vatistas [38].

$$u_\theta^{VV}(r) = \frac{\Gamma}{2\pi R}\cdot\frac{r/R}{\left[1 + (r/R)^{2n}\right]^{\frac{1}{n}}}, \tag{5}$$

which equals to Kaufmann vortex for $n = 1$, and it converges to Rankine vortex for $n \to \infty$.

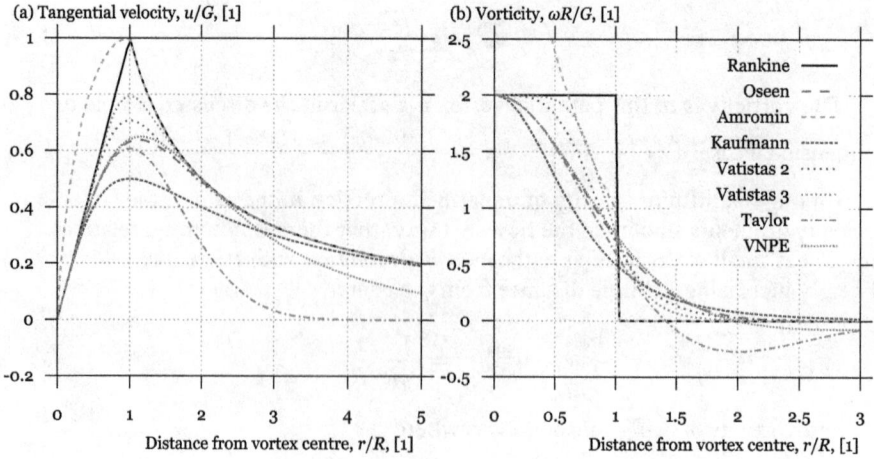

Figure 1.
(a) The tangential velocity profile of discussed vortex models. The velocity is normalized by the circumferential velocity $G = \Gamma/2\pi R$, the distance is normalized by the vortex core radius R. (b) The profiles of theoretical vorticity of the discussed vortex models. The mainstream of vortex models converges to hyperbolic velocity decay $\sim 1/r$ for large r thus having zero vorticity in the far field. Faster velocity decay is redeemed by a skirt of opposite vorticity around the core, see curves denoted "Taylor" and "VNPE."

All the vortex models mentioned up to here display the hyperbolic decrease of tangential velocity with distance, $u_\theta \sim r^{-1}$, see **Figure 1**. Such a vortex has infinite energy! No matter, which profile is found in its core. Let us integrate the kinetic energy of the orbiting fluid since some distance A large enough to eliminate the different core descriptions:

$$E = \int_A^\infty \frac{1}{2} u^2(r) \cdot 2\pi r \cdot dr = \frac{1}{2}\left(\frac{\Gamma}{2\pi R}\right)^2 2\pi \int_A^\infty \left(\frac{R}{r}\right)^2 r\, dr = \pi \left(\frac{\Gamma}{2\pi R}\right)^2 R^2 \left[\ln \frac{r}{R}\right]_A^\infty = \infty$$

(6)

independently on A or other details near the core. This divergence is often solved by declaring some maximum size B of the area influenced by the vortex, which is the size of the experimental cell. It could be the size of a laboratory or the circumference of a planet. Anyway, it is an arbitrary parameter the total energy depends on. It signifies that the distant regions have the same weight as the near regions. This is a very uncomfortable property.

A faster decay of tangential velocity can be found in the Taylor vortex [39].

$$u_\theta^{TV}(r) = \frac{\Gamma}{2\pi R} \cdot \frac{r}{R} \cdot e^{-\frac{1}{2}(r/R)^2},$$

(7)

which is obtained as the first order of *Laguerre polynomials* solution for vorticity, whose zero-th order is the already mentioned Oseen vortex. The detailed mathematics can be found in the book [40], specifically, the Section 6.2.1., and it will be not reproduced here. The velocity decays quite fast and thus the energy converges

$$E = \frac{1}{2}\int_0^\infty u^2(r) 2\pi r\, dr = \pi G^2 R^2 \int x^3 e^{-x^2} dx = \frac{\pi}{2} G^2 R^2,$$

(8)

where[1] $G = \frac{\Gamma}{2\pi R}$ represents an characterisitc vortex core velocity and $x = \frac{r}{R}$ is the dimensionless distance.

The other hand of faster velocity decay is a skirt of vorticity opposite to that in the core. Let us apply the vorticity operator in cylindrical coordinates

$$\omega_z(r) = \left(\nabla \times \vec{u}\right)_z = \frac{1}{r}\left(\frac{\partial r u_\theta}{\partial r} - \frac{\partial u_r}{\partial \theta}\right) = \frac{1}{r}\frac{\partial}{\partial r}rG\frac{r}{R}e^{-r^2/2R^2} = \frac{G}{R}e^{-r^2/2R^2}\left(2 - \frac{r^2}{R^2}\right), \quad (9)$$

which changes the sign at $r/R = \sqrt{2}$, the opposite vorticity value reaches its maximum at $r/R = 2$, and then it decays toward zero. The skirt of opposite vorticity is a property of any profile with tangential velocity decay faster than $1/r$, as the profile $1/r$ is the limiting case for zero vorticity, see **Figure 1**.

2. Experimental setup and methods

2.1 Particle Image Velocimetry

The experimental data were obtained by using the standard method of Particle Image Velocimetry (PIV) [41], which is already a standard tool in hydrodynamic research in air or water and even in superfluid helium [42] as well as in high-speed applications [43]. Contrary to the pressure probes, hot wire anemometry, or laser Doppler anemometry, the result of this method is an instantaneous two-dimensional velocity field [44],[2] which opens the exploration of the turbulent flows topology [8, 45, 46]. It is based on the optical observation of small particles [47] carried by the fluid. The particles are illuminated by a double-pulsed laser in order to capture their movement during the time between pulses. There exists even a time-resolved PIV, which uses fast laser and camera, and thus it is able to capture the temporal development and measure, e.g., the temporal spectra [48]. Our system at the University of West Bohemia in Pilsen belongs to the slow ones with repeating frequency 7.4Hz; therefore, only the statistical properties can be studied with quite good spatial resolution 64×64 grid points sampled on a 4Mpix ($2048 \times 2048\text{pix}^2$) camera images.

2.2 Observed velocity profiles

Let us look at the experimental data. To get the velocity profile of a vortex, it is needed to know vortex parameters: position, radius, and circulation or the effective circumferential velocity $G = \Gamma/2\pi R$ respectively. The listed parameters are results of the fitting procedure; however, the fitting procedure needs to use some vortex model to minimize its energy and, therefore, the result is already a product of the used vortex model. To avoid this back-loop effect, only the *prefit* is used. This function is explained later; it uses the spatial distribution of modified Q-invariant of the velocity gradient tensor and does not need any vortex model explicitly. The vortex velocity profile is obtained as an ensemble average of measured velocity profiles across the vortex; the spatial coordinate is normalized by the vortex radius R and the velocity by the vortex circumferential velocity G. The standard deviation of such ensemble is displayed as a shadow area in **Figures 2–6**.

---

[1] The integral $\int x^3 e^{-x^2} dx$ is solved by substituting $\xi = x^2$, then $d\xi = 2x dx$ and integral is $\frac{1}{2}\int \xi e^{-\xi} d\xi$; per-partes we get $-\frac{1}{2}\left(\xi e^{-\xi} - \int e^{-\xi} d\xi\right) = -\frac{1}{2}e^{-\xi}(\xi + 1)$, i.e. $-\frac{1}{2}e^{-x^2}(x^2 + 1)$.

[2] To obtain the full velocity gradient tensor, Regunath and coworkers had to use 2 laser systems of different colors with slightly shifted planes [44].

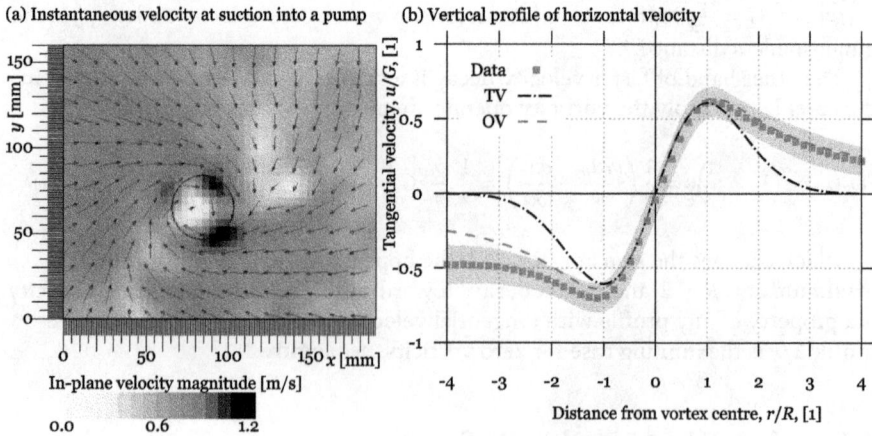

(a) Instantaneous velocity at suction into a pump   (b) Vertical profile of horizontal velocity

In-plane velocity magnitude [m/s]

0.0   0.6   1.2

Distance from vortex centre, r/R, [1]

Figure 2.
(a) Example of instantaneous velocity field measured at a plane perpendicular to the axis of a pump sucking the fluid out. There is found a single vortex in each snapshot, no fitting is used, only the prefit based on the Q criterion. (b) The velocity profile across the found vortices. The profile line is adapted to the position of each vortex, it is rescaled to each vortex radius and each velocity is normalized by the circumferential velocity G, then it is averaged; standard deviation of the ensemble is displayed as a transparent shadow. The theoretical profiles of Taylor vortex (dash-dotted) and Oseen vortex (dashed) are displayed as well.

Figure 2(a) shows the experimental data measured in a plane perpendicular to *suction vortex* formed near the inlet to a pump pumping water from reservoir. The flow field is dominated by this single vortex, which has strong divergence component, it slightly moves around the center, and other parameters vary as well. **Figure 2(b)** shows that the vortex profile in this case roughly follows the Oseen vortex; however, in one direction (toward the left-hand side in the figure), the velocity decays even slower than the Oseen vortex model predicts. This is caused probably by the reservoir geometry. This data were measured by prof. Uruba.
    Figure 3 shows the secondary flow in a corner (bottom and left edge of the figure) of a channel, the main flow is perpendicular to the measured plane. The

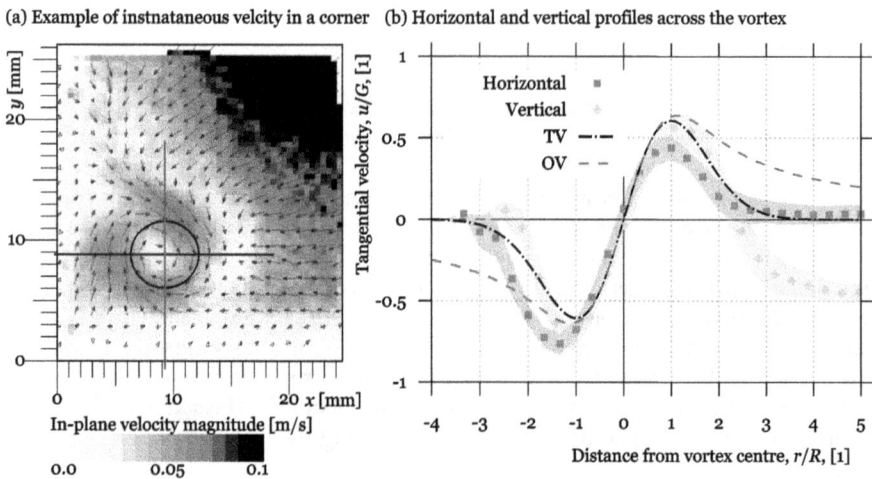

(a) Example of instantaneous velcity in a corner   (b) Horizontal and vertical profiles across the vortex

In-plane velocity magnitude [m/s]

0.0   0.05   0.1

Distance from vortex centre, r/R, [1]

Figure 3.
Example of instantaneous velocity field measured in a plane perpendicular to the main flow through a channel close to the channel corner (physical corner is in the bottom left corner of the figure). When the boundary layers are laminar, a single symmetry-breaking vortex is formed close to the channel corner. (b) Vertical profiles of horizontal velocity and horizontal profiles of vertical velocity across the vortex.

(a) Example of instantaneous velocity past a grid   (b) Vertical profile of horizontal velocity

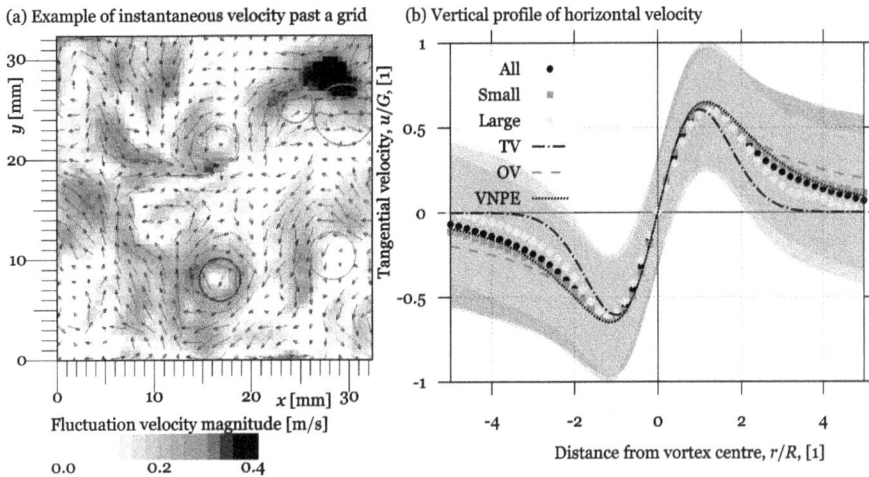

Fluctuation velocity magnitude [m/s]

0.0   0.2   0.4

Figure 4.
(a) Example of instantaneous fluctuating velocity field (fluctuating in respect to the instantaneous field average) with five vortices in each snapshot, no fitting is used. The radius of the vortex is calculated by using the number of grid points contributing to the corresponding patch of Q-criterion. (b) Profiles of the found vortices. The set denoted small contains approx. 10% of vortices with the smallest radius, the large set consists of approx 10% of the largest vortices. The theoretical profiles of Taylor vortex (dash-dotted), Oseen vortex (dashed), and a vortex with non-potential envelope (dotted) are displayed as well.

displayed vortex forms, when the boundary layers are laminar, this vortex spontaneously brakes the symmetry, and it leads to faster transition to turbulence of the boundary layers at higher velocities. More details about this measurement can be found in our previous publications [49, 50]. In this case, the vortex profile is pushed toward the Taylor profile, which is caused by the presence of solid wall and thus zero velocity at the left and bottom side. In the upper direction, there is observed even an overshoot of the profile caused by the stream supplying the vortex from the central flow.

Figure 4 shows the turbulent flow behind a grid; the distance is 200mm, i.e., $12.8M$, M is the mesh parameter, Reynolds number is $3.1 \cdot 10^3$. The main flow points from left to right and the convective velocity component is subtracted. More details about this experiment can be found in our previous publication [51]. In this case, the flow field is not dominated by a single vortex; instead, there are more vortices of similar level. A consequence is that the standard deviation is much more massive than in the previous cases. The averaged profile displays velocity decay faster than the potential envelope ($\sim r^{-1}$), but not as fast as the Taylor vortex model (7).

A similar velocity profile can be seen in a very different case—the jet flow, see **Figure 5**, which shows the data measured in a plane perpendicular to the jet axis at distance of one nozzle diameter past the nozzle. The jet-generating device misses the flow straightener; therefore the jet core contains turbulence originating in the fan; more details can be found in our conference contribution [52]. The vortices prefitted within the jet core (depicted by the blue rectangle in **Figure 5**) display slightly faster velocity decay than the vortices elsewhere, i.e., mainly in the shear layer.

A highly turbulent flow emerges in the steam turbines; the data measured in a model axial air turbine are shown in **Figure 6**. Here, the strong advection in the axial direction (from left to right in the figure) is subtracted, the rest shows a wide horizontal strip of lower turbulence, which is caused by the rotor jet (fluid passing the interblade channel), this structure overlays a less apparent structure of wakes past rotor wheel, which display as strips of wilder flow in top-bottom direction.

(a) Velocity perpendicular to a jet, $x/D = 1.0$ (b) Vertical profile of horizontal velocity

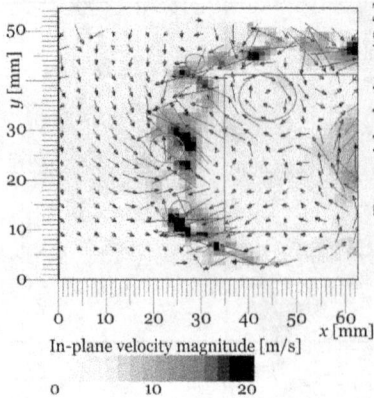

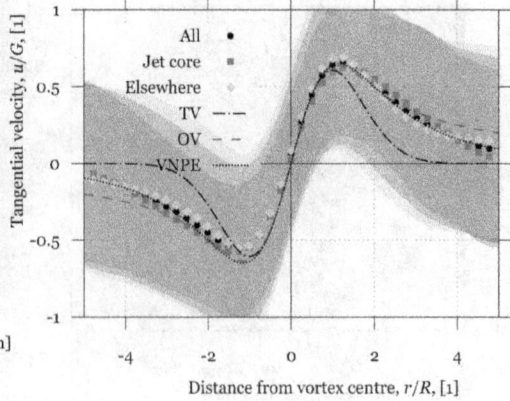

In-plane velocity magnitude [m/s]

Distance from vortex centre, r/R, [1]

Figure 5.
(a) Example of instantaneous in-plane velocity field perpendicular to a turbulent jet axis with five vortices in each snapshot, no fitting is used. (b) Profiles of the found vortices. The set is approximately separated into vortices in the jet core and elsewhere according to the blue rectangle in panel (a).

(a) Fluctuations past single turbine stage   (b) Axial and tangential profiles of radial vortices

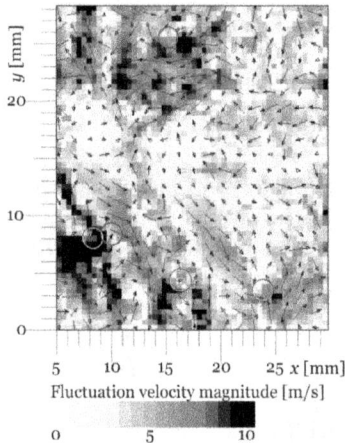

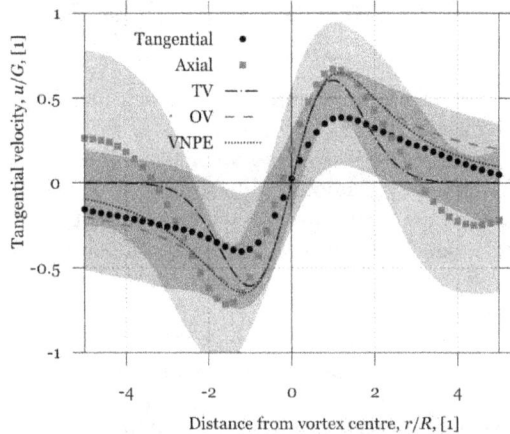

Fluctuation velocity magnitude [m/s]

Distance from vortex centre, r/R, [1]

Figure 6.
(a) Example of instantaneous velocity field in the axial × tangential plane inside an axial turbine past the first stage (stator + rotor); the convective velocity in axial direction (from left to right) is subtracted. (b) Profiles of the found vortices. The profiles in axial direction display strong overshoot caused by the pattern of wakes past rotor wheel (such wakes pass the field of view from top to bottom with small left-right drift as the rotor wheel rotates from bottom to top in the field of view perspective).

This pattern would be better apparent in an averaged image, but here the instantaneous field is shown. More detailed description of this interesting flow can be found in our article [53]. The vortices in this case display a strong asymmetry—in tangential direction (up-down in the figure), their peak velocity is significantly smaller than the peak velocity in axial direction (left-right). In the axial direction, there is a strong "overshoot" of velocity decay, which is caused by the alternating velocity pattern.

The observation made in very different cases does not support the generally accepted hypothesis of potential envelope around the vortex. This envelope forms, when there is only single vortex dominating the flow (see **Figure 2** with single suction vortex), in the case of turbulent field consisting of multiple vortices, the velocity decay is faster, but not as fast as in the Taylor vortex model (7). Therefore we venture to offer another model of vortex with non-potential envelope

$$u_\theta^{\text{VNPE}}(r) = \frac{\Gamma}{2\pi R} \cdot \frac{r/R}{\left(1 + \frac{1}{4}(r/R)^2\right)^2} \qquad (10)$$

which decays as r^{-3} at large r and at smaller r, it roughly follows the Oseen vortex model, see **Figure 1**. It is important to note that this vortex is not a solution of Navier-Stokes equations! It is based on the observations only, and there is no any theoretical argument for it.

Similarly as the Taylor vortex, this model displays a skirt of opposite vorticity as well. The vorticity profile is

$$\omega_z(r) = G \frac{2 - \frac{1}{2}(r/R)^2}{\left(1 + \frac{1}{4}(r/R)^2\right)^3}. \qquad (11)$$

It reaches zero at the distance

$$0 = 2 - \frac{1}{4}\left(\frac{r_0}{R}\right)^2 \Rightarrow \frac{r_0}{R} = 2, \qquad (12)$$

see **Figure 1**; since this distance, the vorticity approaches zero from opposite direction as $\sim r^{-4}$ for large r. The energy of this vortex model is finite even in unbounded domain:

$$E = \frac{1}{2}\int_0^\infty \left[G\frac{r/R}{\left(1 + \frac{(r/R)^2}{4}\right)^2}\right]^2 \cdot 2\pi r dr = \frac{4}{3}\pi G^2 R^2. \qquad (13)$$

2.3 Vortex prefit

Any general fitting algorithm falls into some local minimum. This minimum does not need to be the really wanted results, it can be just a small dimple in a wall of huge valley. To avoid this effect, one can (i) modify the fitting algorithm to see larger surroundings of the point, e.g., by using *simulated annealing* [54, 55], or (ii) just start close to the result. The second possibility solves another small issue—the starting point of the fitting algorithm. In the single particular case solved here, it means to find a peak of appropriate scalar variable, which would signify the presence of vortex.

The prefit is sketched in **Figure 7**: the starting point is the spatial scalar field of $\sqrt{Q_d} \cdot \text{sgn}\,\omega$, where $\text{sgn}\,\omega = \frac{\omega}{|\omega|}$ is the sign of vorticity. Q_d is the Q-invariant with subtracted divergence. Alternatively, any scalar with sparse non-zero values could be used (i.e., not simply the vorticity). Then the separated patches of non-zero signal are detected, see panel (c) of **Figure 6**. The vortex is built up by using the most energetic patch (label 3 in **Figure 7(c)**); the vortex position is the center of mass of the patch, the vortex radius $R = \sqrt{n/\pi}$, where n is the number of points of the patch (note that the unit of R is the grid point). The circumferential velocity G is calculated as the average of tangential projection of the measured velocities at eight locations around the vortex in the distance R from its center (crosses in **Figure 7(d)**).

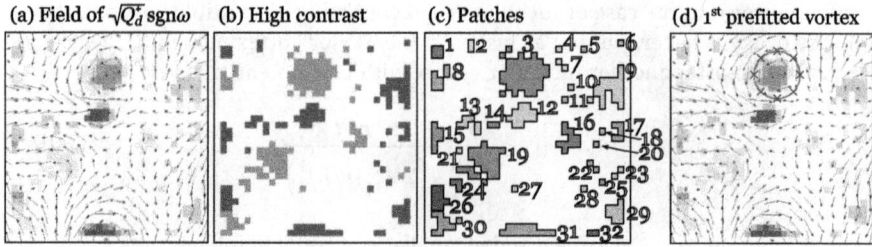

| (a) Field of $\sqrt{Q_d^+}$ sgnω | (b) High contrast | (c) Patches | (d) 1ˢᵗ prefitted vortex |

Figure 7.

Steps of prefit: (a) calculate the scalar field of $\sqrt{Q_d^+}$ sgn ω, which produces a lot of zeros, thus the areas with non-zero value (panel (b)) are sparse and thus the percolation does not occur. (c) The patches or individual continuous areas of non-zero $\sqrt{Q_d^+}$ sgn ω are identified and the one with the largest energy (label three in this case) serves for generating the prefitted vortex. Note that the patches numbered 26 and 30 would merge if the sign of local vorticity was not used to separate the opposite orientations. (d) The prefitted vortex with estimated radius from the number of points in the patch; the crosses around the vortex show the eight positions of velocity estimation.

The just described procedure does not use the vortex model; therefore it is suitable for velocity profile estimation as has been done in the previous section. On the other hand, vortex parameters are only estimated; therefore the fitting is needed to adapt the vortex parameters to the actual velocity field.

2.4 Vortex fitting

A single vortex is described by four fitting parameters: the position x and y, core radius R, and circumferential velocity G, which is easier to use than the circulation $\Gamma = 2\pi G$. Of course, this set of parameters describes only the cases, when the vortex tube crosses the measured plane perpendicularly; other angles might produce deformation from the ideal circular shape. But, as John von Neumann said: *With four parameters I can fit an elephant, and with five I can make him wiggle his trunk* [56]. Therefore, it is preferred to avoid using too many fitting parameters; the listed set is considered to be a minimum. This issue will be a true challenge in the case of instantaneous volumetric data in the future.

The used fitting algorithm is called *Amoeba* [55] or *Downhill simplex method* or *Nelder-Mead* by its inventors [57]. The energy of residual velocity field is calculated for each variant serving as a score. Here comes the need of specific vortex model discussed earlier. The algorithm selects single movement from a closed set of movements in the parameter space in order to keep away from areas with high residual energy and converging to some local minimum. The algorithm is in much more detail described in the book [55].

Once the energy residual of a single vortex reaches local minimum, this vortex is subtracted from the velocity field. Then the entire procedure is repeated by using the residual velocity field as the input.

Figure 8 shows the results of fitting a single instantaneous velocity field by depicted number of vortices. It is clearly visible that the energy decreases as the field is approximated by more and more vortices. It can be seen as a kind of decomposition, although its effectivity is poor in comparison with pure mathematical approaches, e.g., the Proper Orthogonal Decomposition [8, 9]. On the other hand, it describes the fluctuating velocity field by using objects with clear physical interpretation, while the physical interpretation of POD modes is not straightforward [58]. Still, a question remains here: whatever the found vortices are real. To be specific, in **Figure 8**, a large vortex can be seen even in the first set, the core of this

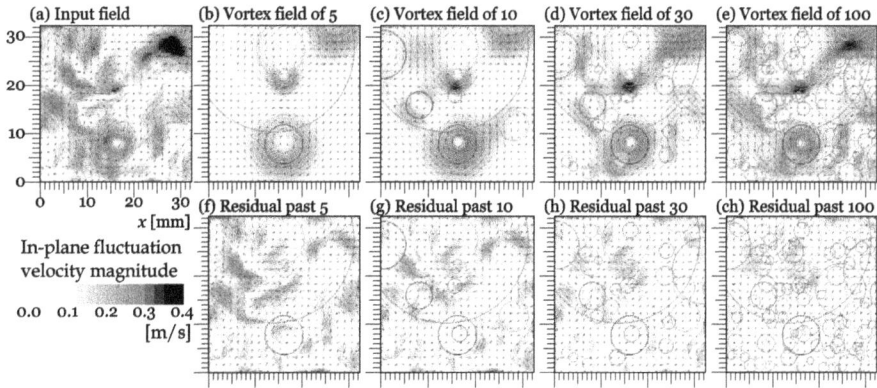

Figure 8.
Vortices fitted in a single instantaneous velocity field measured past a grid; the same example field as in previous figures. (a) The input velocity field, (b, c, d, e) the velocity field calculated from theoretical vortex profiles. (f, g, h, ch) The residual field, i.e., input field minus the field of found vortices.

vortex spans out of the field of view, thus no one knows, if the vortex was still there in the case that areas were measured. For example, a simple advective motion can be explained by a pair of huge vortices up and down the measured area. Of course, that is unphysical. As the number of vortices increases, even smaller and smaller vortices are added converging to a situation, that each single noise vector is described by a single vortex. This limit is unphysical as well, but where is the boundary?

Figure 9 shows the decrease of effectivity of this procedure—as the number of vortices increases, there remains structures less and less similar to a vortex in the instantaneous velocity field. While the first vortex typically covers around 10% of the energy of input fluctuating velocity field (in this case). The convergence of the energy of the rest gets slower, and it becomes to be quite ineffective to describe 75% of the fluctuations by the simple vortices described here. The parameters of found vortices develop as well, see **Figure 10,** which shows the probability density functions of vortex core radii and circumferential velocities. The vortices found later are typically smaller and have smaller circumferential velocity (the positive and negative values count together in the logarithmic plot of **Figure 10(c)** and **(d)**).

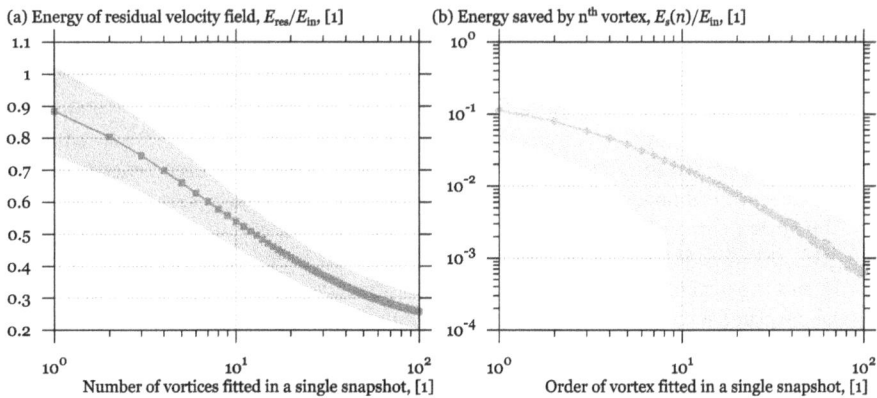

Figure 9.
(a) Energy of the residual velocity field after subtracting the nth vortex as a function of number of vortices. The area represents the standard deviation of the ensemble of 1476 snapshots. (b) The energy "saved" by nth vortex, again, the area represents the standard deviation of the ensemble.

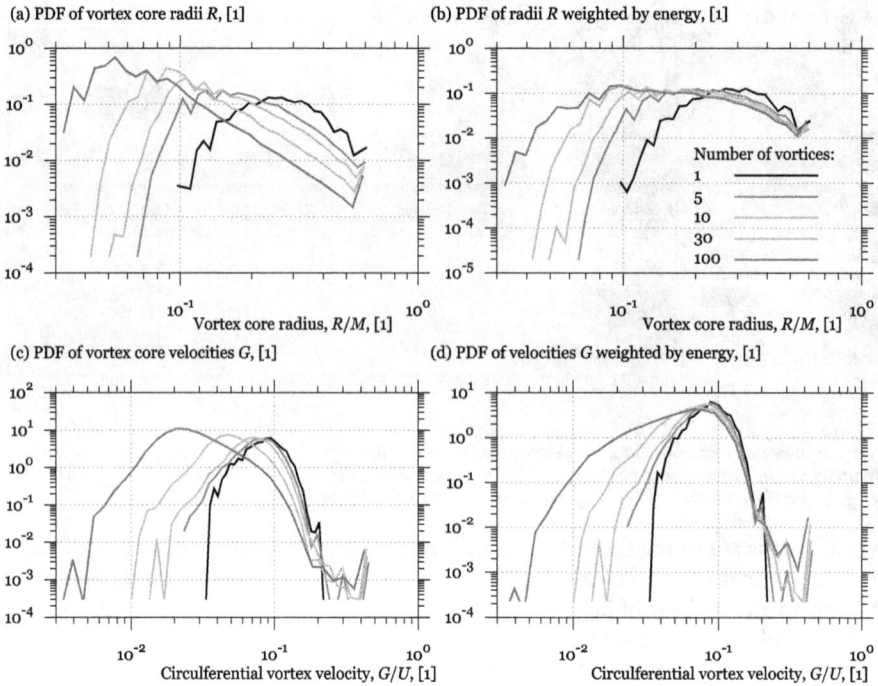

Figure 10.
Probability density functions (PDFs) of core radii R (a and b) and circumferential velocities G (c and d) for several number of vortices searched in a single instantaneous velocity field—Black: Single vortex in each field, maroon: 5 vortices, yellow: 10 vortices, green: 30 vortices and blue: 100 vortices. The PDF can be weighted by the number of vortices (i.e., one vortex, one vote) or by the energy, panels (b and d).

3. Results

The aim of this chapter is mainly to describe the ideas of the developed algorithm. The results and their physical interpretation need more effort in the future. In this section, some ways of result analysis are shown based on the distribution study. As a example case, the grid turbulence is selected, because this is a deeply explored canonical case, see [59] and many more experimental data, e.g., in [4, 60–65].

When exploring the vortices in grid turbulence in dependence on the distance behind the grid or on the Reynolds number, the *effectivity* of vortex fitting remains almost constant. **Figure 11** shows, that the maximum population is around 1 in all cases. There is slow decrease of the number of vortices with low effectivity (i.e., structures with large radius or large velocity causing large theoretical energy, $E_t \sim R^2 G^2$, which do not correspond to the energy saved), and there exist cases with saved energy larger than the theoretical one; however, this distribution decreases much faster.

The vortex core radii in **Figure 12(a)** do not seem to depend on the distance x past the grid, although it is known that the characteristic turbulent length scales (Kolmogorov and Integral one) typically increase with distance. At the lowest distance, there can be observed a weak wavening of the distribution. The distribution dependence on Reynolds number is weak as well (**Figure 12(b)**), although a very fine change of vortex core radii scaling at radii larger maximal population. Honestly speaking, the similarity of the distributions is suspicious, and it has to be proven in the future that the shape of the radii distribution is not affected by the measurement spatial resolution (the studied datasets have all the same spatial resolution).

(a) PDF of vortex effectivity, [100 %], different distance *x*   (b) PDF of vortex effectivity, [100 %], different Reynolds

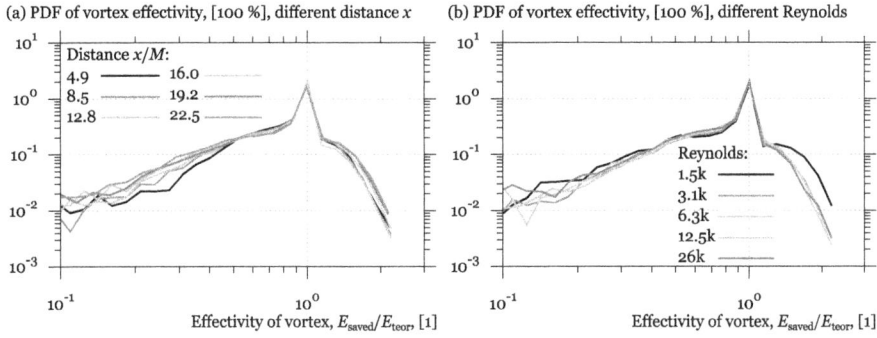

Figure 11.
Probability density functions (PDFs) of the vortex effectivity, i.e., the ratio of energy saved by the probed vortex and the theoretical energy of the vortex. Left panel (a) shows the data at different distances behind the grid, the mesh-based Reynolds number is $3.1 \cdot 10^3$; panel (b) shows the data at different Reynolds number, the distance $x/M = 12.8$. The "k" in the legend plays for $\cdot 10^3$.

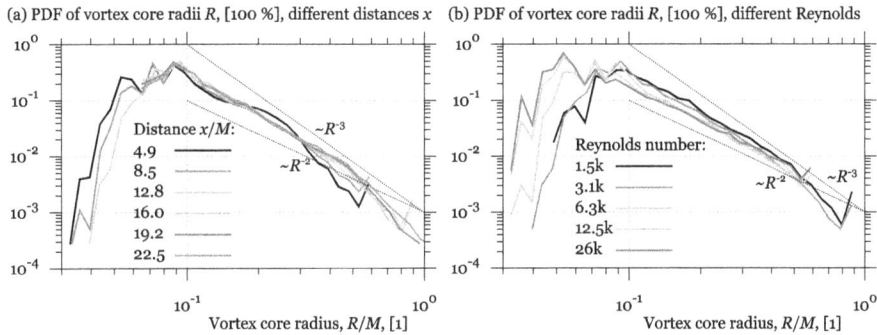

(a) PDF of vortex core radii *R*, [100 %], different distances *x*   (b) PDF of vortex core radii *R*, [100 %], different Reynolds

Figure 12.
Probability density functions (PDFs) of core radii R found in fields of view in several distances past the grid, panel (a); and at several Reynolds numbers, panel (b). The dotted lines highlight scalings of R^{-3} and R^{-2} the observed data lie in between. It seems that the scaling exponent slightly decreases with Re.

The circumferential velocities $G = \Gamma/2\pi R$ of the vortices move toward smaller values with increasing distance, see **Figure 13(a)**. This effect is clearly caused by the decreasing turbulence intensity [51] as the vortices are searched within the fluctuating velocity field. **Figure 13(b)** shows that the velocity normalized by the wind tunnel velocity of maximum population increases. At lower velocities, the PDF decrease with increasing *G* displays two regimes, first it decreases slower, then faster, while at higher velocities, only the fast decay is observable. It has to be mentioned, in the light of observations in **Figure 9**, that this effect can be caused by the number of fitted vortices, which do not need to be appropriate for the actual datasets. It is quite difficult to distinguish the effects of the method and the physical phenomena.

The distance to nearest other vortex seems to be unaffected by the grid distance and flow velocity, see **Figure 14**. But the absolute values of the nearest vortex cannot have some physical sense, as this quantity is the first one dependent on the number of searched vortices, thus the vortex density. But the non-changing shape of this distribution suggests that there is nothing like evolution pattern of vortices or vortex lattice.

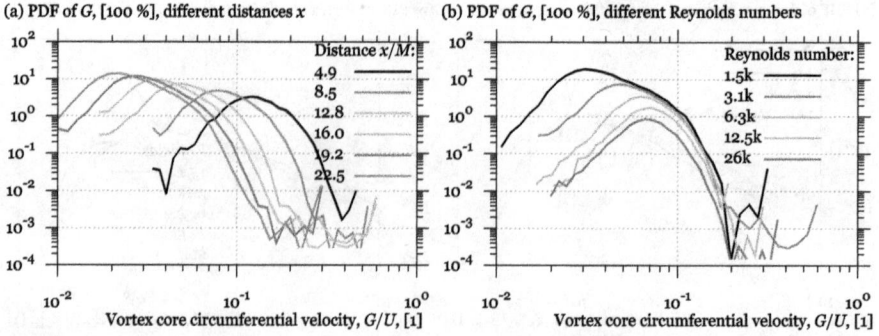

Figure 13.
Probability density functions (PDF) of core circumferential velocities G for several distances past the grid, panel (a); and several wind tunnel velocities changing the Reynolds number, panel (b).

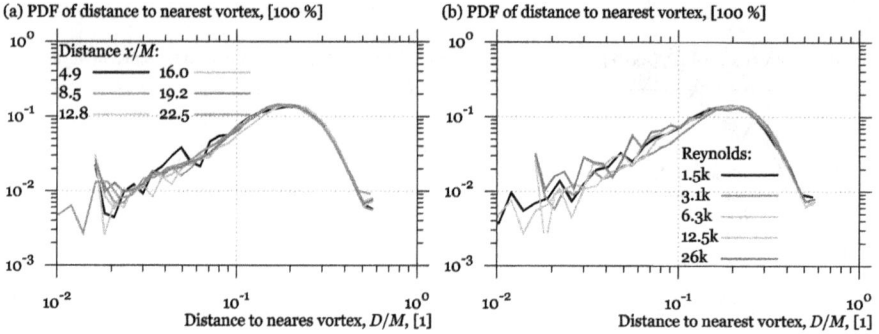

Figure 14.
Probability density functions (PDF) of the distance to nearest vortex for several distances past the grid, panel (a); and several Reynolds numbers, panel (b).

4. Conclusions

The turbulent flows consist of many interacting vortices of all scales, which all together self-organize being responsible for the statistical properties of turbulence. In this contribution, the algorithm for detection of individual vortices via direct fitting of measured velocity field has been presented. It has been shown via the zero-th step of fitting that the velocity profile of vortex in turbulent flow decreases faster than the generally accepted models suggest. This is advantageous, because the energy of vortex with velocity decrease faster than $1/r$ converges. On the other hand, it has a "skirt" of vorticity opposite to the center one. The vortices found in grid turbulence display average radius decreasing with distance and Reynolds number, while the scaling at larger R seems to not depend on those parameters. The effective circumferential velocity $G = \Gamma/2\pi R$ decreases with distance and increases with Reynolds number (faster than expected linear). The algorithm is still under development and mainly the physical interpretation of the results needs more work in the future studying and comparing results of different flow cases.

Acknowledgements

Great thanks belong to my colleagues: Václav Uruba and Vitalii Yanovych.

The research was supported from ERDF under project LoStr No. CZ.02.1.01/0.0/0.0/16_026/0008389. The publication was supported by project CZ.02.2.69/0.0/0.0/18_054/0014627.

Nomenclature

E	energy; E_{in} energy of the input velocity field, E_s energy saved by removing fitted vortices, E_{res} energy of the velocity field after removing vortices, E_t theoretical energy of the vortex
G	vortex core circumferencial velocity, $G = \frac{\Gamma}{2\pi R}$
M	mesh parameter of the grid, i.e., the distance of the rods
R	vortex core radius
Q	invariant of velocity gradient tensor, in 2D: $Q = \partial_x u \partial_y v - \partial_y u \partial_x v$
Q_d	Q invariant without the divergence, in 2D: $Q_d = -\partial_x u \partial_y v - \partial_y u \partial_x v - (\partial_x u)^2 - (\partial_y v)^2$
u	instantaneous velocity, $\vec{u}$ vector, u_θ tangential velocity,
v	velocity component perpendicular to u in 2D.
Γ	vortex circulation
ω	vorticity $\vec{\omega} = \nabla \times \vec{u}$

Abbreviations

AV	Amromin vortex $u_\theta(r) = G \cdot \frac{r}{R} \cdot \left(1 - \ln \frac{r}{R}\right)$ for $r < R$ and potential vortex outside
KV	Kaufmann (often reported as Scully) vortex, $u_\theta(r) = G \cdot \frac{r/R}{1+(r/R)^2}$
OV	Oseen Vortex, $u_\theta(r) = G \cdot \frac{R}{r}\left(1 - \exp\left(-\frac{r^2}{R^2}\right)\right)$
PDF	probability density function
PIV	Particle Image Velocimetry
PV	potential vortex, $u_\theta(r) = G \cdot \frac{R}{r}$
SBR	solid-body rotation, $u_\theta(r) = G \cdot \frac{r}{R}$
TV	Taylor vortex, $u_\theta(r) = G \cdot \frac{r}{R} \exp\left(-\frac{r^2}{2R^2}\right)$
VNPE	vortex with non-potential envelope, $u_\theta(r) = G \cdot \frac{r/R}{\left[1+\left(\frac{r}{2R}\right)^2\right]^2}$
VV	Vatistas vortex system, $u_\theta(r) = G \cdot \frac{r/R}{\left[1+(r/R)^{2n}\right]^{1/n}}$

Author details

Daniel Duda
University of West Bohemia in Pilsen, Pilsen, Czech Republic

*Address all correspondence to: dudad@kke.zcu.cz

IntechOpen

References

[1] Uriel Frisch and Andre Nikolaevich Kolmogorov. Turbulence: The Legacy of AN Kolmogorov. Cambridge: Cambridge University Press; 1995

[2] La Mantia M, Švančara P, Duda D, Skrbek L. Small-scale universality of particle dynamics in quantum turbulence. Physical Review B. 2016;**94**(18)

[3] Tabeling P, Zocchi G, Belin F, Maurer J, Willaime H. Probability density functions, skewness, and flatness in large reynolds number turbulence. Physical Review E. 1996;**53**:1613-1621

[4] Burgoin M et al. Investigation of the small-scale statistics of turbulence in the modane s1ma wind tunnel. CEAS Aeronautical Journal. 2018;**9**(2):269-281

[5] Azevedo R, Roja-Solórzano LR, Leal JB. Turbulent structures, integral length scale and turbulent kinetic energy (tke) dissipation rate in compound channel flow. Flow Measurement and Instrumentation. 2017;**57**:10-19

[6] Schulz-DuBois EO, Rehberg I. Structure function in lieu of correlation function. Applied Physics. 1981;**24**: 323-329

[7] Kirkpatrick S, Gelatt CD, Vecchi MP. Optimization by simulated annealing. Science. 1983;**220**:671-680

[8] Uruba V. Decomposition methods for a piv data analysis with application to a boundary layer separation dynamics. *Transactions of the VŠB – Technical University of Ostrava Mechanical Series.* 2010;**56**:157-162

[9] Uruba V. Energy and entropy in turbulence decompositions. Entropy. 2019;**21**(2):124

[10] Richardson LF. Atmospheric diffusion shown on a distance-neighbour graph. Proceedings of the Royal Society A. 1926;**110**:709-737

[11] Fiedler HE. Coherent structures in turbulent flows. Progress in Aerospace Sciences. 1988;**25**(3):231-269

[12] Barkley D, Song B, Mukund V, Lemoult G, Avila M, Hof B. The rise of fully turbulent flow. Nature. 2015; **526**(7574):550-553

[13] Valente PC, Vassilicos JC. The decay of turbulence generated by a class of multiscale grids. Journal of Fluid Mechanics. 2011;**687**:300-340

[14] Barenghi CF, Skrbek L, Sreenivasan KR. Introduction to quantum turbulence. Proceedings of National Academy of Sciences of the United States of America. 2014;**111**: 4647-4652

[15] Vinen WF. An introduction to quantum turbulence. Journal of Low Temperature Physics. 2006;**145**(1–4): 7-24

[16] Fonda E, Meichle DP, Ouellette NT, Hormoz S, Lathrop DP. Direct observation of Kelvin waves excited by quantized vortex reconnection. Proceedings of the National Academy of Sciences. 2014;**111**(Supplement_1): 4707-4710

[17] Helmholtz H. Über integrale der hydrodynamischen gleichungen, welche den wirbelbewegungen entsprechen. Journal für die reine und angewandte Mathematik. 1858;**55**:25-55

[18] Baggaley AW, Laurie J, Barenghi CF. Vortex-density fluctuations, energy spectra, and vortical regions in superfluid turbulence. Physical Review Letters. 2012;**109**(20):205304

[19] La Mantia M, Duda D, Rotter M, Skrbek L. Velocity statistics in quantum turbulence. In: Procedia IUTAM. Vol. 9. 2013. pp. 79-85

[20] Staicu AD. Intermittency in Turbulence. Eidhoven: University of Technology Eidhoven; 2002

[21] Duda D, Švančara P, La Mantia M, Rotter M, Skrbek L. Visualization of viscous and quantum flows of liquid he 4 due to an oscillating cylinder of rectangular cross section. Physical Review B—Condensed Matter and Materials Physics. 2015;**92**(6)

[22] Duda D, Yanovych V, Uruba V. An experimental study of turbulent mixing in channel flow past a grid. PRO. 2020; **8**(11):1-17

[23] Hänninen R, Baggaley AW. Vortex filament method as a tool for computational visualization of quantum turbulence. Proceedings of the National Academy of Sciences of the United States of America. 2014;**111**(Suppl. 1): 4667-4674

[24] Varga E, Babuin S, V. S. L'vov, A. Pomyalov, and L. Skrbek. Transition to quantum turbulence and streamwise inhomogeneity of vortex tangle in thermal counterflow. Journal of Low Temperature Physics. 2017;**187**(5–6):531-537

[25] De Gregorio F, Visingardi A, Iuso G. An experimental-numerical investigation of the wake structure of a hovering rotor by piv combined with a γ_2 vortex detection criterion. Energies. 2021;**14**(9):2613

[26] Marchevsky IK, Shcheglov GA, Dergachev SA. On the algorithms for vortex element evolution modelling in 3d fully lagrangian vortex loops method. In Topical Problems of Fluid Mechanics. 2020;**2020**:152-159

[27] Kaufmann W. Über die ausbreitung kreiszylindrischer wirbel in zähen (viskosen) flüssigkeiten. Ingenieur-Archiv. 1962;**31**(1):1-9

[28] Ben-Gida H, Liberzon A, Gurka R. A stratified wake of a hydrofoil accelerating from rest. Experimental Thermal and Fluid Science. 2016;**70**: 366-380, 105374 p.

[29] Jeong J, Hussain F. On the identification of a vortex. Journal of Fluid Mechanics. 1995;**285**:69-94

[30] Dergachev SA, Marchevsky IK, Shcheglov GA. Flow simulation around 3d bodies by using lagrangian vortex loops method with boundary condition satisfaction with respect to tangential velocity components. Aerospace Science and Technology. 2019;**94**:105374

[31] Graftieaux L, Michard M, Grosjean N. Combining PIV, POD and vortex identification algorithms for the study of unsteady turbulent swirling flows. Measurement Science and Technology. 2001;**12**(9):1422-1429

[32] Koschatzky V, Moore PD, Westerweel J, Scarano F, Boersma BJ. High speed piv applied to aerodynamic noise investigation. Experiments in Fluids. 2011;**50**(4):863-876

[33] Maciel Y, Robitaille M, Rahgozar S. A method for characterizing cross-sections of vortices in turbulent flows. International Journal of Heat and Fluid Flow. 2012;**37**:177-188, 118108 p.

[34] Amromin E. Analysis of vortex core in steady turbulent flow. Physics of Fluids. 2007;**19**:118108

[35] Keshavarzi A, Melville B, Ball J. Three-dimensional analysis of coherent turbulent flow structure around a single circular bridge pier. Environmental Fluid Mechanics. 2014;**14**:821-847

[36] Scully MP, Sullivan JP. Helicopter rotor wake geometry and airloads and

development of laser doppler velocimeter for use in helicopter rotor wakes. In: Technical Report. MIT; 1972. Available from: http://citeseerx.ist.psu.edu/viewdoc/summary?doi=10.1.1.982.2439

[37] Bhagwat MJ, Leishman JG. Generalized viscous vortex model for application to free-vortex wake and aeroacoustic calculations. Annual Forum Proceedings-American Helicopter Society. 2002;**58**:2042-2057

[38] Vatistas GH, Kozel V, Mih WC. A simpler model for concentrated vortices. Experiments in Fluids. 1991;**11**(1):73-76

[39] Taylor GI. On the Dissipation of Eddies. ACA/R&M-598. London: H.M. Stationery Office; 1918

[40] Wu JZ, Ma HY, Zhou MD. Vorticity and Vortex Dynamics. Berlin Heidelberg New York: Springer; 2006

[41] Tropea C, Yarin A, Foss JF. Springer Handbook of Experimental Fluid Mechanics. Berlin Heidelberg: Springer; 2007

[42] La Mantia M, Skrbek L. Quantum turbulence visualized by particle dynamics. Physical Review B—Condensed Matter and Materials Physics. 2014;**90**(1):1-7

[43] Kurian T, Fransson JHM. Grid-generated turbulence revisited. Fluid Dynamics Research. 2009;**41**(2):021403

[44] Regunath GS, Zimmerman WB, Tesař V, Hewakandamby BN. Experimental investigation of helicity in turbulent swirling jet using dual-plane dye laser PIV technique. Experiments in Fluids. 2008;**45**(6):973-986

[45] Agrawal A. Measurement of spectrum with particle image velocimetry. Experiments in Fluids. 2005;**39**(5):836-840

[46] Romano GP. Large and small scales in a turbulent orifice round jet: Reynolds number effects and departures from isotropy. International Journal of Heat and Fluid Flow. 2020;**83**:108571

[47] Jašková D, Kotek M, Horálek R, Horčička J, Kopecký V. Ehd sprays as a seeding agens for piv system measurements. In: ILASS – Europe 2010, 23rd Annual Conference on Liquid Atomization and Spray Systems; 23 September 2010; Brno, Czech Republic. p. 2010

[48] Jiang MT, Law AW-K, Lai ACH. Turbulence characteristics of 45 inclined dense jets. Environmental Fluid Mechanics. 2018;**19**:1-28

[49] Bém J, Duda D, Kovařík J, Yanovych V, Uruba V. Visualization of secondary flow in a corner of a channel. In: AIP Conference Proceedings. Vol. 2189. 2019. pp. 020003-1-020003-6

[50] Duda D, Jelínek T, Milčák P, Němec M, Uruba V, Yanovych V, et al. Experimental investigation of the unsteady stator/rotor wake characteristics downstream of an axial air turbine. International Journal of Turbomachinery, Propulsion and Power. 2021;**6**(3)

[51] Duda D. Preliminary piv measurement of an air jet. AIP Conference Proceedings. 2018;**2047**:020001

[52] Duda D, Bém J, Yanovych V, Pavlíček P, Uruba V. Secondary flow of second kind in a short channel observed by piv. European Journal of Mechanics, B/Fluids. 2020;**79**:444-453

[53] Duda D, La Mantia M, Skrbek L. Streaming flow due to a quartz tuning fork oscillating in normal and superfluid he 4. Physical Review B. 2017;**96**(2):024519

[54] Kolář V. Vortex identification: New requirements and limitations. International Journal of Heat and Fluid Flow. 2007;**28**(4):638-652

[55] Press WH, Teukolsky SA, Vetterling WT, Flannery BP. Numerical Recipes 3rd Edition: The Art of Scientific Computing. Cambridge: Cambridge University Press; 2007

[56] Meyer J, Khairy K, Howard J. Drawing an elephant with four complex parameters. American Journal of Physics. 2010;**78**:648-649

[57] Nelder JA, Mead R. A simplex method for function minimization. The Computer Journal. 1965;**7**(4): 308-313

[58] Uruba V, Hladík O, Jonáš P. Dynamics of secondary vortices in turbulent channel flow. Journal of Physics: Conference Series. 2011;**318**: 062021

[59] Kuzmina K, Marchevsky I, Soldatova I. The high-accuracy numerical scheme for the boundary integral equation solution in 2d lagrangian vortex method with semi-analytical vortex elements contribution accounting. In: Topical Problems of Fluid Mechanics 2020. Prague: Czech Academy of Sciences; 2020. pp. 122-129

[60] Comte-Bellot G, Corrsin S. The use of a contraction to improve the isotropy of grid-generated turbulence. Journal of Fluid Mechanics. 1966;**25**:657-682

[61] Wierciński Z, Grzelak J. The decay power law in turbulence. Transactions of the Institute of Fluid-flow Machinery. 2015;**130**:93-107

[62] Jonáš P, Mazur O, Uruba V. On the receptivity of the by-pass transition to the length scale of the outer stream turbulence. European Journal of Mechanics, B/Fluids. 2000;**19**(5): 707-722

[63] Mohamed MS, LaRue JC. The decay power law in grid-generated turbulence. Journal of Fluid Mechanics. 1990;**219**: 195-214

[64] Roach PE. The generation of nearly isotropic turbulence by means of grids. International Journal of Heat and Fluid Flow. 1987;**8**:82-92

[65] Warhaft Z, Lumley JL. An experimental study of the decay of temperature fluctuations in grid generated turbulence. Journal of Fluid Mechanics. 1978;**88**:659-684

Chapter 4

Vortex Dynamics in Complex Fluids

Naoto Ohmura, Hayato Masuda and Steven Wang

Abstract

The present chapter provides an overview of vortex dynamics in complex fluids by taking examples of Taylor vortex flow. As complex fluids, non-Newtonian fluid is taken up. The effects of these complex fluids on the dynamic behavior of vortex flow fields are discussed. When a non-Newtonian shear flow is used in Taylor vortex flow, an anomalous flow instability is observed, which also affects heat and mass transfer characteristics. Hence, the effect of shear-thinning on vortex dynamics including heat transfer is mainly referred. This chapter also refers to the concept of new vortex dynamics for chemical process intensification technologies that apply these unique vortex dynamics in complex fluids in Conclusions.

Keywords: Taylor vortex flow, complex fluid, non-Newtonian fluid, heat transfer, process intensification

1. Introduction

Historically, innovative processes have been created using organized vortices. For example, in Japan, Kiyomasa Kato, a Sengoku daimyo (Japanese territorial lord in the Sengoku period) in Kumamoto Prefecture, made a canal (called "hanaguri canal") with a partition (baffle) having a semicircular hole at the bottom as shown in **Figure 1**. The flow velocity of the water flowing through the hole in the lower part of the partition increases due to the effect of the contraction of the flow, and a strong circulating vortex is formed in the water channel divided by the partition. By intensifying the flow in the canal, water can be supplied to about 95 ha of land in nine villages in the downstream without piling up volcanic ash or earth and sand, and the harvest has increased about three times. Based on this idea by Kiyomasa Kato, in order to solve the particle sedimentation problem in oscillatory baffled reactors (OBR) which is one of the hopeful process intensification techniques, our group [1] succeeded in preventing the particle sedimentation to the bottom of the reactor and obtaining extremely monodispersed particles in a calcium carbonate crystallization process by changing from a normal baffle with a hole in the center to a snout-type baffle as shown in **Figure 2**.

In addition, the function of vortex flow is not only to intensify the previously noticed transport phenomena such as mixing, heat transfer, and mass transfer, but also to have a new function that has not been previously noticed, such as classification and separation of particles. Ohmura et al. [2] found that particles with different

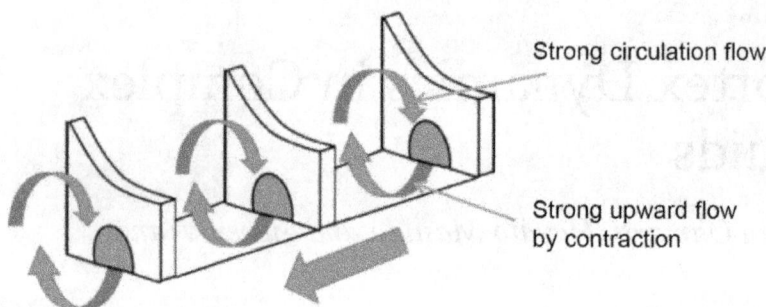

Figure 1.
Schematic of Hanaguri Canal.

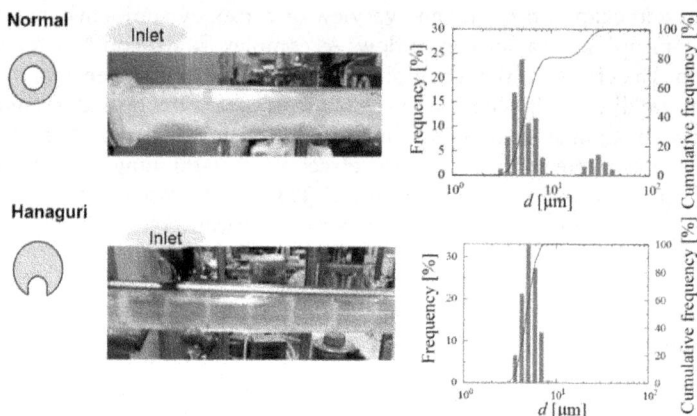

Figure 2.
Comparison of performance of an oscillatory baffled crystallizer between using normal and Hanaguri-type baffles.

sizes move on different streamlines within a Taylor cell and proposed that this could be applied to a particle classification device. Kim et al. [3] applied this idea to a continuous crystallizer and proposed a device for granulating particles of different sizes while classifying them. Wang et al. [4] also proposed a novel solid–liquid separation system that breaks the conventional stereotype of mixing equipment by applying the particle clustering phenomenon in isolated mixing regions in stirring tanks. In this way, vortices with a systematic structure have very attractive properties, such as solid accumulation, mixing and reaction enhancement, particle classification, and mass transport. If we can understand the characteristics of this organized vortex structure and manipulate it freely, we may be able to develop innovative chemical processes.

In many industrial processes, such as chemical, food, and mineral processes, the fluids handled are not only simple homogeneous Newtonian fluids, but also often complex fluids, such as non-Newtonian fluids, multi-phase fluids with highly dispersed phases, and viscoelastic fluids. Therefore, in order to apply the new "vortex dynamics" currently being constructed to process intensification technologies and implement it in society, it is necessary to develop the concept of new "vortex dynamics" from simple fluids to complex fluids. According to the abovementioned background, the present chapter provides an overview of vortex dynamics in complex fluids by taking examples of Taylor vortex flow.

2. Vortex dynamics with non-Newtonian fluids

A non-Newtonian fluid property causes a multiple fluid motion. These motions are quite interesting from fundamental and practical viewpoints. Especially, in vortex flow systems, fluid elements experience curved streamlines. In polymeric fluid systems, the polymer molecule chain does not line along curved stream lines, and consequently, hoop stress in a normal direction occurs. As a result, coupling normal stresses and curved streamlines causes elastic instabilities [5]. These insta-bilities are observed in various flows, e.g., Poiseuille flow [6], microchannel flow [7], and swirling flow [8]. Many polymeric fluids show not only viscoelastic behav-ior but also shear-thinning behavior. The shear-thinning property causes the vis-cosity distribution accompanied by the shear-rate distribution in the fluid system. Coelho and Pinho [9] showed that the shear-thinning affects the flow transition of vortex shedding in a cylinder flow. Ascanio et al. [10] reported that the mixing process of shear-thinning fluids under a time-periodic flow field is different from that of Newtonian fluid. Thus, vortex dynamics in non-Newtonian fluid systems is far from complete.

To investigate the effect of non-Newtonian property on vortex dynamics in more detail, many researchers have been utilizing Taylor–Couette flow, which is one of the most canonical flow systems in fluid mechanics, with non-Newtonian fluids [11–14]. Taylor–Couette flow is the flow between coaxial cylinders with the inner one rotating. This flow shows a cascade transition from laminar Couette flow to fully turbulent wavy vortex flow with the increase in circumferential Reynolds number (Re). When the value of Re exceeds the critical Re (Re_{cr}), Taylor vortex flow firstly appears. As mentioned above, many researchers have been studied the Tay-lor–Couette flow with non-Newtonian fluids. For example, Muller et al. [11] and Larson et al. [12] revealed that the elastic instability occurs in Taylor–Couette flow and organized flow modes based on Deborah number (De), which the ratio of a characteristic relaxation time of the fluid to a characteristic residence time in the flow geometry [5]. **Figure 3** shows laminar Taylor–Couette flow with Newtonian (40 wt% glycerol aqueous solution) and viscoelastic fluid (0.75 wt% sodium polyacrylate aqueous solution).

The flow pattern was visualized by adding a small amount of Kalliroscope AQ-1000 flakes. As shown in **Figure 3**, the cellular structure of Taylor vortices seems to be complicated in the viscoelastic fluid even at the relatively low Re. The detailed mechanism is found in their papers [11–14]. Other interesting point is an

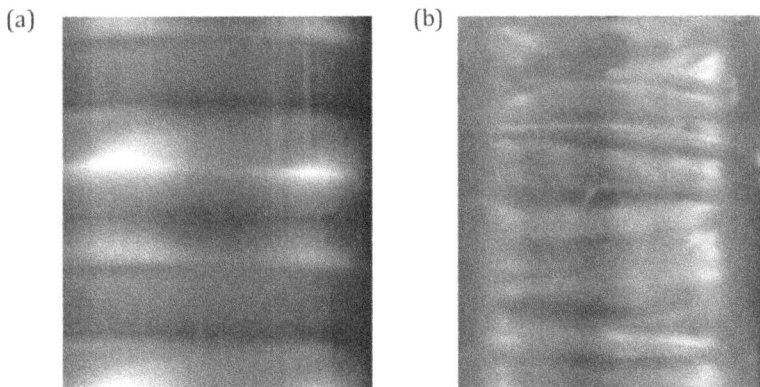

Figure 3.
Flow visualization: (a) Newtonian fluid at Re = 212 and (b) viscoelastic fluid at Re$_{eff}$ = 218.

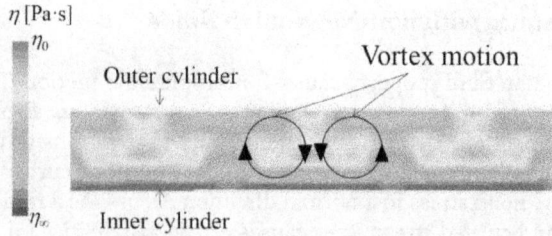

Figure 4.
Viscosity distribution in the annular space obtained by numerical simulation [15]. The fluid was assumed to be a shear-thinning fluid.

enlarged vortex structure by shear-thinning property. Escudier et al. [15] found that the cellular vortex is axially stretched and the vortex eye (the location of zero axial velocity in the vortex interior) is radially shifted toward the center body.

However, the first Taylor–Couette instability has not been fully understood yet in non-Newtonian fluid systems. One of the reasons is the discrepancy between Re_{cr} reported by several researchers for non-Newtonian fluids. Alibenyahia et al. [16] reviewed the discrepancy; Jastrebski et al. [17] reported Re_{cr} decreased with the shear-thinning property, on the other hand, Caton et al. [18] found the opposite tendency. Actually, this discrepancy is explained by the difference in how to define the effective Reynolds number, Re_{eff}, in their papers. In non-Newtonian fluids, how to define Re is quite complicated because the viscosity locally varies as shown in **Figure 4** [19]. Practically, Re_{eff} based on the effective viscosity in the system should be discussed. Several researchers have been trying to define more rational Re_{eff} in various flow systems, e.g., rising bubble flow in shear-thickening fluid [20], Rayleigh–Bénard convection with shear-thinning fluids [21], and non-Newtonian fluid flow past a circular cylinder [22].

We previously proposed a new definition of Re_{eff} based on the effective viscosity (η_{eff}), which is obtained by numerical simulation. η_{eff} is calculated by averaging the locally distributed viscosity using a weight of dissipation function as follows [23]:

$$\eta_{eff} = \sum_{i=1}^{N} \dot{\gamma}_i^2 \eta_i \Delta V_i / \sum_{i=1}^{N} \dot{\gamma}_i^2 \Delta V_i, \tag{1}$$

where N is the total mesh number, η_i [Pa·s] is the local viscosity, $\dot{\gamma}_i$ [1/s] is the local shear rate, and ΔV_i [m^3] is the local volume for each cell. It should be noted that η_{eff} is obtained using numerical simulation. The computational domain is shown in **Figure 5**. The governing equations are as follows:

$$\nabla \cdot \mathbf{u} = 0, \tag{2}$$

$$\frac{\partial \mathbf{u}}{\partial t} + (\mathbf{u} \cdot \nabla)\mathbf{u} = -\frac{\nabla p}{\rho} + \frac{1}{\rho} \nabla \cdot (2\eta \mathbf{D}) + \mathbf{g}, \tag{3}$$

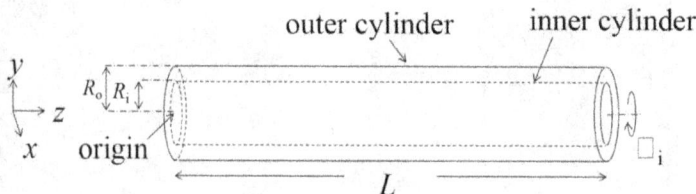

Figure 5.
Computational domain [22]. R_i and R_o are the radii of inner and outer cylinders, respectively.

where $\mathbf{u}$ [m/s] is the velocity, p [Pa] is the pressure, ρ [kg/m^3] is the density, η [Pa·s] is the viscosity depending on the shear rate, $\mathbf{D}$ (= $(\nabla \mathbf{u} + \nabla \mathbf{u}^T)$ / 2) [1/s] is the rate of deformation tensor, $\mathbf{g}$ [m/s^2] is the gravitational acceleration. The rheological property is characterized by Carreau model as follows [24]:

$$\eta = \eta_0 \left[1 + (\beta \cdot \dot{\gamma})^2 \right]^{(n-1)/2},$$
(4)

where η_0 [Pa·s] is the zero shear-rate viscosity, $\dot{\gamma}$ [1/s] is the shear rate, β [s] is the characteristic time, and n [−] is the power index, which indicates the slope of decreasing viscosity with shear rate. In the case of $n < 1$, the fluid shows the shear-thinning behavior. The detailed information of numerical procedure is written in our paper [23].

Figure 6 shows the critical value of Re_{eff} for various shear-thinning fluids as a function of gap ratio R_i / R_o. The theoretical Re_{cr} for Newtonian fluids derived by Taylor [25] was denoted by the dashed line in **Figure 6**. It is found that the critical Re_{eff} for shear-thinning fluids was in agreement with the theoretical value at R_i / $R_o > 0.7$. Thus, Re_{eff} defined based on η_{eff} by Eq. (1) is rational as a practical basis. The effect of shear-thinning property on the vortex structure is also interesting from the viewpoint of fluid dynamics. **Figure 7** shows the number of pairs of Taylor

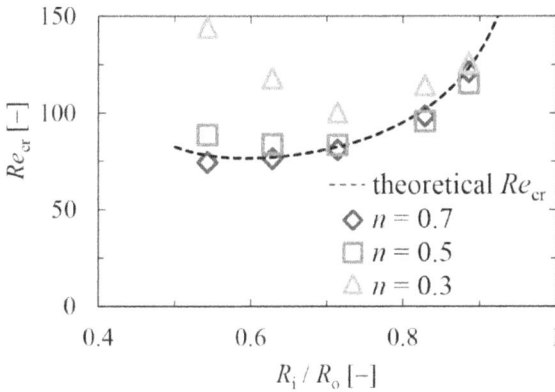

Figure 6.
Re_{cr} for various shear thinning [18].

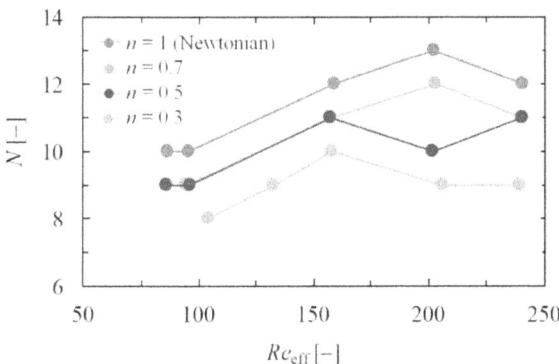

Figure 7.
Variation in the number of pairs of Taylor vortices [23].

cells, N, as a function of Re_{eff} at the aspect ratio Γ = 20 [26]. In all fluid systems, N tended to increase with Re_{eff}. This tendency agrees with reports by other researchers [27]. Furthermore, the shear-thinning property seems to make Taylor cells large because N decreases with the shear-thinning property at the same degree of Re_{eff}. This tendency was remarkable in the case of n = 0.3. This means that the shear-thinning property axially enlarges Taylor cells. Although the detailed mechanism of enlarging Taylor cells is under consideration, it will be clarified by numerical simulation of development process of Taylor vortices.

We also introduce heat transfer characteristics of Taylor–Couette flow with shear-thinning fluids. In addition to Eqs. (2) and (3), energy equation was solved:

$$\frac{\partial}{\partial t}\left(\rho C_p\right) + \nabla \cdot \left(\rho C_p T \mathbf{u}\right) = \nabla \cdot \left(\kappa \nabla T\right), \tag{5}$$

where C_p [J/kg·K] is the specific heat capacity, T [K] is the temperature, and κ [J/m·s·K] is the thermal conductivity. **Figure 8** shows the axial variation in the local Nusselt number, Nu_L, at the surface of the outer cylinder at Re_{eff} = 158 [26]. The Nu_L at the surface of the outer cylinder was calculated as follows:

$$Nu_L = \frac{2hd}{\kappa}, \tag{6}$$

where h is a local heat transfer coefficient. As clearly shown in **Figure 6**, Nu_L decreases with the increase in the shear-thinning property. This decrease is explained by increasing the thickness of velocity boundary layer for shear-thinning fluid systems (**Figure 9**). Generally speaking, it is said that the shear-thinning property improves heat transfer performance at same Re [28, 29]. This is because the viscosity reduction by the shear-thinning property is not adequately reflected in Re used in papers. In other words, the actual flow condition is underestimated in the case of shear-thinning fluids. Thus, the heat transfer performance is not accurately compared between Newtonian and shear-thinning fluids unless Re_{eff} is used for representation of flow condition.

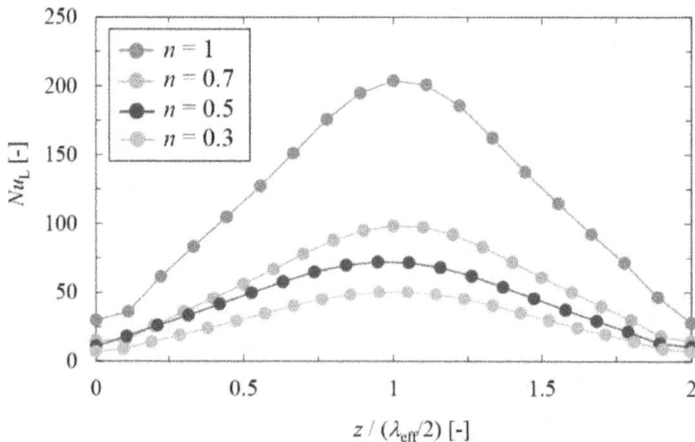

Figure 8.
Axial variation in the local Nusselt number (Nu$_L$) along the surface of the outer cylinder at Re$_{eff}$ = 158 [23]. λ_{eff} is the wavelength of Taylor cells.

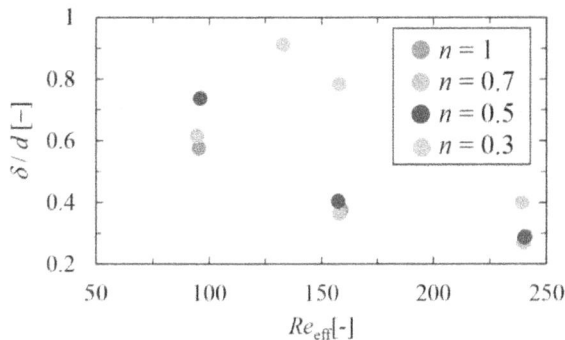

Figure 9.
Dependence of the dimensionless thickness of the velocity boundary layer [23].

3. Conclusions

In this section, we mainly refer the effect of shear-thinning on vortex dynamics including heat transfer. However, the viscoelastic property further complicates vortex dynamics as shown in **Figure 3**. In the future, vortex dynamics and transport phenomena in viscoelastic fluid systems should be investigated in more detail. In this case, it is considered to be important to construct a mathematical model by multi-scale analysis focusing on the interaction among scales of microstructure (molecular structure of polymers, micelles, particles, etc.), mesostructure (entanglement of polymer, particle aggregation, etc.), and macrostructure (vortex flow) of complicated fluid. For example, when a polymer solution flows in a micro channel having a sharp contraction part, an unsteady vortex called viscoelastic turbulence is generated in a corner part of the contraction part at higher Weissenberg number [30]. When the scale of the microchannel becomes small, the scale of the flow can be compared with the scale of the polymer. Since the influence of the elasticity derived from the deformation of the polymer itself on the flow becomes large, there is a possibility that the dynamic characteristics of the vortex generated in the contraction part can be controlled by the channel shape. In order to construct a methodology of controlling the viscoelastic vortex, a multi-scale simulation combined with molecular dynamics and computational fluid dynamics may be important.

As this viscoelastic vortex example shows, the field in which the vortex occurs affects the characteristics of the vortex. In the case of a Taylor vortex flow system, for example, the structure and dynamic characteristics of the vortices largely depend on the surface properties. It has been reported that heat transfer is enhanced by processing regular unevenness in the circumferential direction on the outer cylinder surface [31]. In the case of conical Taylor vortex flow, our previous work [32] successfully reproduced the phenomenon that the vortices move upward spontaneously under specific conditions by numerical analysis, and it was found that mass transfer was enhanced in polymer fluid system. In this way, it is possible to control the characteristics of the vortex flow by a structurally organized (having low entropy or fractal) nonuniform field rather than simply a random (high-entropy) nonuniform field. Therefore, in order to systematize a new vortex dynamics for freely manipulating vortices, it is necessary to quantitatively express the heterogeneity by introducing the concept of entropy and fractal and to clarify the relationship between the structure of the field and the characteristics of vortices.

Acknowledgements

This work was supported by the KAKENHI Grant-in-Aid for Scientific Research (A) JP18H03853 and the Fostering Joint International Research (B) JP19KK0127.

Conflict of interest

The authors declare no conflict of interest.

Author details

Naoto Ohmura[1]*, Hayato Masuda[2] and Steven Wang[3]

1 Kobe University, Kobe, Japan

2 Osaka City University, Osaka, Japan

3 Hong Kong City University, Hong Kong, China

*Address all correspondence to: ohmura@kobe-u.ac.jp

IntechOpen

References

[1] Amano K, Horie T, Ohmura N, Watabe Y. Analysis of fluid dynamics in an oscillatory baffled reactor for continuous crystallization. In: Proceedings of the 6th International Workshop on Process Intensification (IWPI2018). Taipei: National Taiwan University; 2018. pp. 106-107

[2] Ohmura N, Suemasu T, Asamura Y. Particle classification in Taylor vortex flow with an axial flow. Journal of Physics: Conference Series. 2005;**14**: 64-71. DOI: 10.1088/1742-6596/14/009

[3] Kim JS, Kim DH, Gu B, Kim DY, Yang DR. Simulation of Taylor-Couette reactor for particle classification using CFD. Journal of Crystal Growth. 2013; **373**:106-110. DOI: 10.1016/j. crysgro.2012.12.006

[4] Wang S, Metcalfe G, Stewart RL, Wu J, Ohmura N, Feng X, et al. Solid-liquid separation by particle-flow-instability. Energy & Environmental Science. 2014;**7**:3982-3988. DOI: 10.1039/c4ee02841d

[5] Pakdel P, McKinley GH. Elastic instability and curved streamlines. Physical Review Letters. 1996;**77**: 2459-2462. DOI: 10.1103/ PhysRevLett.77.2459

[6] Joo YL, Shaqfeh ESG. Viscoelastic Poiseuille flow through a curved channel: A new elastic instability. Physics of Fluid A: Fluid Dynamics. 1991;**3**:2043-2046. DOI: 10.1063/1.857886

[7] Hong SO, Cooper-White JJ, Kim JM. Inertio-elastic mixing in a straight microchannel with side wells. Applied Physics Letters. 2016;**108**:014103. DOI: 10.1063/1.4939552

[8] Yao G, Yang H, Zhao J, Wen D. Experimental study on flow and heat transfer enhancement by elastic instability in swirling flow. International Journal of Thermal Sciences. 2020;**157**: 106504. DOI: 10.1016/j. ijthermalsci.2020.106504

[9] Coelho PM, Pinho F. Vortex shedding in cylinder flow of shear-thinning fluids. I. Identification and demarcation of flow regime. Journal of Non-Newtonian Fluid Mechanics. 2003; **110**:110143-110176. DOI: 10.1016/ S0377-0257(03)00007-7

[10] Ascanio G, Foucault S, Tanguy PA. Time-periodic mixing of shear-thinning fluids. Chemical Engineering Research and Design. 2004;**82**:1199-1203. DOI: 10.1205/cerd.82.9.1199.44155

[11] Muller SJ, Larson RG, Shaqfeh ESG. A purely elastic transition in Taylor-Couette flow. Rheologica Acta. 1989;**28**: 499-503. DOI: 10.1007/BF01332920

[12] Larson R, Shaqfeh E, Muller S. A purely elastic instability in Taylor–Couette flow. Journal of Fluid Mechanics. 1990;**218**:573-600. DOI: 10.1017/S0022112090001124

[13] Groisman A, Steinberg V. Couette-Taylor flow in a dilute polymer solution. Physical Review Letters. 1996;**77**: 1480-1483. DOI: 10.1103/ PhysRevLett.77.1480

[14] Cagney N, Lacassagne T, Balabani S. Taylor–Couette flow of polymer solutions with shear-thinning and viscoelastic rheology. Journal of Fluid Mechanics. 2020;**905**:A28. DOI: 10.1017/jfm.2020.701

[15] Escudier MP, Gouldson IW, Jones DM. Taylor vortices in Newtonian and shear-thinning liquids. Proceedings of The Royal Society A. 1995;**449**: 155-176. DOI: 10.1098/rspa.1995.0037

[16] Alibenyahia B, Lemaitre C, Nouar C, Ait-Messaoudene. Revisiting

the stability of circular Couette flow of shear-thinning fluids. Journal of Non-Newtonian Fluid Mechanics. 2012; **183-184**:37-51. DOI: 10.1016/j.jnnfm.2012.06.002

[17] Jastrzębski M, Zaidani HA, Wroński S. Stability of Couette flow of liquids with power law viscosity. Rheologica Acta. 1992;**31**:264-273. DOI: 10.1007/BF00366505

[18] Caton F. Linear stability of circular Couette flow of inelastic viscoplastic fluids. Journal of Non-Newtonian Fluid Mechanics. 2006;**134**:148-154. DOI: 10.1016/j.jnnfm.2006.02.003

[19] Masuda H, Horie T, Hubacz R, Ohmura N, Shimoyamada N. Process development of starch hydrolysis using mixing characteristics of Taylor vortices. Bioscience, Biotechnology, and Biochemistry. 2017;**81**:755-761. DOI: 10.1080/09168451.2017.1282806

[20] Ohta M, Kimura S, Furukawa T, Yoshida Y, Sussman M. Numerical simulations of a bubble rising through a shear-thickening fluid. Journal of Chemical Engineering of Japan. 2012;**45**: 713-720. DOI: 10.1252/jcej.12we041

[21] Jenny M, Plaut E, Briard A. Numerical study of subcritical Rayleigh–Bénard convection rolls in strongly shear-thinning Carreau fluids. Journal of Non-Newtonian Fluid Mechanics. 2015;**219**:19-34. DOI: 10.1016/j.jnnfm.2015.03.002

[22] Ohta M, Toyooka T, Matsukuma. Numerical simulations of Carreau-model fluid flows past a circular cylinder. Asia-Pacific Journal of Chemical Engineering. 2020;**15**:e2527. DOI: 10.1002/apj.2527

[23] Masuda H, Horie T, Hubacz R, Ohta M, Ohmura N. Prediction of onset of Taylor-Couette instability for shear-thinning fluids. Rheologica Acta. 2017;**56**: 73-84. DOI: 10.1007/s00397-016-0987-7

[24] Carreau PJ. Rheological equations from molecular network theories. Transactions of the Society of Rheology. 1972;**16**:99-127. DOI: 10.1122/1.549276

[25] Taylor GI. Stability of a viscous liquid contained between two rotating cylinders. Philosophical Transactions of the Royal Society A. 1923;**223**:289-343. DOI: 10.1098/rsta.1923.0008

[26] Masuda H, Shimoyamada M, Ohmura N. Heat transfer characteristics of Taylor vortex flow with shear-thinning fluids. International Journal of Heat and Mass Transfer. 2019;**130**: 274-281. DOI: 10.1016/j.ijheatmasstransfer.2018.10.095

[27] Neitzel GP. Numerical computation of time-dependent Taylor-vortex flows in finite-length geometries. Journal of Fluid Mechanics. 1984;**141**:51-66. DOI: 10.1017/S0022112084000732

[28] Izadpanah E, Rabiee MB, Sadeghi H, Talebi S. Effect of rotating and oscillating blade on the heat transfer enhancement of non-Newtonian fluid flow in a channel. Applied Thermal Engineering. 2017;**113**:1277-1282. DOI: 10.1016/j.applthermaleng.2016.11.124

[29] Crespí-Llorens D, Vicente P, Viedma A. Experimental study of heat transfer to non-Newtonian fluids inside a scraped surface heat exchanger using a generalization method. International Journal of Heat and Mass Transfer. 2018;**118**:75-87. DOI: 10.1016/j.ijheatmasstransfer.2017.10.115

[30] Rodd LE, Cooper-White JJ, Boger DV, McKinley GH. Role of the elasticity number in the entry flow of dilute polymer solutions in micro-fabricated contraction geometries. Journal of Non-Newtonian Fluid Mechanics. 2007;**143**:170-191. DOI: 10.1016/j.jnnfm.2007.02.006

[31] Nouri-Borujerdi A, Nakhchi ME. Optimization of the heat transfer

coefficient and pressure drop of Taylor-Couette-Poiseuille flows between an inner rotating cylinder and an outer grooved stationary cylinder. International Journal of Heat and Mass Transfer. 2017;**108**:1449-1459. DOI: 10.1016/j.ijheatmasstransfer.2017.01.014

[32] Masuda H, Iyota H, Ohmura N. Global convection characteristics of conical Taylor–Couette flow with shear-thinning fluids. Chemical Engineering & Technology. 2021;**44**:2049-2055. DOI: 10.1002/ceat.202100236

Chapter 5

Vortex Analysis and Fluid Transport in Time-Dependent Flows

Stefania Espa, Maria Grazia Badas and Simon Cabanes

Abstract

In this contribution, we present a set of procedures developed to identify fluid flow structures and characterize their space-time evolution in time-dependent flows. In particular, we consider two different contests of importance in applied fluid mechanics: 1) large-scale almost 2D atmospheric and oceanic flows and 2) flow inside the left ventricle in the human blood circulation. For both cases, we designed an ad hoc experimental model to reproduce and deeply investigate the considered phenomena. We will focus on the post-processing of high-resolution velocity data sets obtained via laboratory experiments by measuring the flow field using a technique based on image analysis. We show how the proposed methodologies represent a valid tool suitable for extracting the main patterns and quantify fluid transport in complex flows from both Eulerian and Lagrangian perspectives.

Keywords: pattern identification, laboratory experiments, image analysis, rotating turbulence, flow in the left ventricle

1. Introduction

In most of the fluid flows of interest in nature and technology (i.e., geophysical flows, blood flow in the human circulation as well as flows in turbomachinery and around vehicles) the presence of turbulence in normally observed; therefore, their reproducibility and repeatability have always represented a crucial issue. In this regard, it is widely recognized that laboratory experiments represent a valid tool for the simulation and investigation of complex fluid flows under controlled conditions. With the improvement of measuring techniques, the possibility of acquiring huge high-resolution data sets in space and time is continually increasing. It is then fundamental to consider procedures suitable for a proper analysis of these data aimed at the definition and the characterization of the main flow pattern and of their evolution. In this contribution, we consider two examples of different contests of importance in applied fluid mechanics: 1) β-plane turbulence in the framework of large-scale almost 2D atmospheric and oceanic flows and 2) effect of artificial valves on the flow in the left ventricle in the framework of an in vitro model of human blood circulation. In both cases, the complexity of the flow arises from the embedded non-linear phenomena i.e., interaction of structures at different scales, the interplay between vortices waves and turbulence, anisotropy in the energy transfers, and in transport phenomena. Due to chaotic advection, the Lagrangian motion of passive particles can be very complex even in regular, i.e., non-turbulent, flow fields [1] as in the situations here discussed in which we considered almost 2D and time-periodic velocity fields. The chapter is organized as follows. In Section 2, we describe the case studies and the considered experimental apparatus. Theory, its application to the experiments, and the different post-processing methodology are described in Section 3, Section 4 contains some results. We discuss and give our conclusions in Section 5.

2. Material and methods

We provide below the description of the experimental models designed to reproduce: 1) turbulent flows affected by a β-effect, 2) the flow downstream a natural/artificial valve in the left ventricle as well as an overview of the technique used to measure the velocity fields.

2.1 Rotating turbulent flows with a β-effect

In rotating turbulent flows, the latitudinal variation of the Coriolis parameter, the so-called β-effect, may redirect the upward energy flux towards the zonal modes thus inducing the anisotropization of the inverse energy cascade, typically observed in large-scale geophysical flows. Due to the combined effects of planetary rotation, topographical constraints, and fluid stratification, these circulations can be assumed quasi-two-dimensional to the first degree of approximation. Actually, the anisotropic inverse energy cascade represents one of the leading causes for the formation and maintenance of jet-like structures along the zonal direction, the so-called zonation [2–5] observed in the atmospheres of the Giant Planets and in the terrestrial oceans. These environments are characterized by the existence of a banded structure, i.e., eastward and westward zonal flows, as well as by the coexistence of turbulence and waves on all scales [6].

In this contest, in addition to the characteristic scales of 2D turbulence [7] associated with the small-scale forcing k_f and the large-scale friction k_{fr}, two more wavenumbers have to be considered: the Rhines wavenumber k_{Rh} and the transitional wavenumber k_β. The Rhines wavenumber is defined as the scale at which the velocity root-mean-square U_{RMS} is equal to the phase speed of Rossby waves $k_{Rh} = (\beta/2U_{RMS})^{1/2}$ [2] and can be related to the meridional size of the jet. If the flow is continuously forced at small scales and at a constant rate ε, the balance between the eddy characteristic time and the Rossby wave period exists in correspondence of $k_\beta = (\beta^3/\varepsilon)^{1/5}$, i.e., the so-called anisotropic transitional wavenumber which characterizes the threshold of the inverse cascade anisotropization [8]. The ratio between the transitional wavenumber and the Rhines wavenumber provides the non-dimensional number, $R_\beta = k_\beta/k_{Rh}$, known as the zonostrophy index. This represents a key parameter in turbulent flows subjected to a β-effect, since it discerns different flow regimes of the so-called β-plane turbulence. Indeed $R_\beta < 1.5$ pertains to flows with strong large-scale friction (friction dominated regime), in the range $1.5 < R_\beta < 2.5$ a flow shows a transitional behavior, and for $R_\beta > 2.5$ a flow develops within the regime of zonostrophic turbulence [8].

To deeply investigate these features, we carried out several experimental campaigns in a rotating tank facility available at the Hydraulics Laboratory of the Sapienza University of Rome. As reported in previous papers [9, 10], the experimental setup consists of a square tank 1 m in diameter placed on a rotating table whose imposed rotation is counter-clockwise in order to emulate flows in the Northern hemisphere of a planet. To simulate the dynamics associated with the latitudinal variation of the Coriolis parameter in the Polar Regions, we consider the effects induced by the parabolic shape assumed by the free surface of a rotating fluid. In fact, it is represented by a quadratic variation in r, being r the radial distance from the pole and assuming the pole as the reference point (polar β-plane or γ plane approximation) [10, 11]. In this model, the center of the tank (i.e., the point of maximum depression of the fluid surface) represents the pole, while the periphery of the domain corresponds to lower latitudes.

In particular, a local Cartesian frame of reference at the midlatitude of the tank ($r_m = R/2$; where R is the radius of the tank) was considered to evaluate the strength of the β term in each experiment [9]. We run a huge set of experiments by changing the main parameters of the flow, i.e. the rotation rate of the system, the fluid thickness, the amount of energy introduced into the system as well as the forcing characteristics [12–15]. Here, we focus on the analysis of the flow induced by a localized forcing, i.e. the formation of a single eastward/westward jet. To this aim, we consider an electromagnetic forcing obtained with the Lorentz force arising from the interaction of a horizontal electric field and a vertical magnetic field.

We perform a set of runs in which the magnets are located along an arc of latitude in the range $180° < \varphi < 360°$ at a distance $r = 17$ cm from the pole; the considered angular velocity and fluid depth at rest are $\Omega = 3 \mathrm{rads}^{-1}$ and $H_0 = 4$ cm, respectively. To force the flow, we considered the same orientation of polarity chosen such as to introduce an eastward/westward momentum and facilitate the formation of an eastward/westward zonal jet; in fact, the stationary position of the magnets locked the jet's location. In each of these runs, we vary the intensity of the current in the range $2A \leq I \leq 6A$; the forcing was continuously applied for all the duration of the experiments.

2.2 Flow in the left ventricle in the human blood circulation

The overall functionality of the heart pump is strongly related to the intraventricular flow features. Complexity in the ventricular flow is mainly due to fluid-wall interactions and turbulence onset in correspondence of the boundaries, three-dimensionality, and asymmetry in the pattern development. Here, the focus is on the investigation of the flow in the left ventricle (LV) during a cardiac cycle: it consists of an intense jet forming downstream of the mitral valve and in the development of the related coherent structures i.e., a vortex ring, which grows up during the systole, impinges on the ventricle walls and vanishes almost completely during the systole. A deeper analysis of the flow pattern evolution has shown on one hand that the observed flow structure appears to be favorable to ejection through the aortic valve during the systole [16] and on the other hand the mutual relationships between the formation and development of coherent structures in the LV and its functionality. Actually, one of the main reasons for the deviation from physiological conditions is represented by the replacement of the mitral valve with a prosthetic one, which obviously causes deep modifications in the hemodynamics and, consequently, in the associated flow pattern [17–19].

We reproduce in the laboratory the ventricular flow by means of a pulse duplicator widely described in previous papers [19–21], below we summarize its working principle. A flexible, transparent sack made of silicone rubber (wall thickness ~ 0.7 mm) simulates the LV allowing at the same time for the optical access. The model ventricle is fixed on a circular plate, 56 mm in diameter, and connected to a constant-head tank by means of two Plexiglas conduits. Along the outlet (aortic) conduit a check valve was mounted, whereas different types of valves were placed on the inlet (mitral) orifice.

We consider three different scenarios: a) the inlet was designed in order to obtain a uniform velocity profile at the orifice mimicking physiological conditions, b) a monoleaflet (Bjork–Shiley monostrut) in mitral position 3) a bileaflet bicarbon prosthetic valve in mitral position; both valves were 31 mm in nominal diameter. The model of the LV was placed in a rectangular tank with Plexiglas (transparent) walls; its volume changed according to the motion of the piston, placed on the side of the tank. The piston was driven by a linear motor, controlled by means of a speed-feedback servo-control. The motion assigned to the linear motor was tuned to reproduce the volume change by clinical data acquired in vivo by echo-cardiography on a healthy subject [20].

2.3 Measuring technique

Two-dimensional velocity fields are measured by means of an image analysis technique called Feature Tracking, FT [22, 23]. The measurement chain can be summarized in the following steps: 1) identification of a proper measurement plane in the fluid domain; 2) seeding of the working fluid with a passive tracer; 3) illumination of the measurement plane previously identified; 4) image acquisition; 5) image pre-processing of the acquired images; 6) particle detection and temporal tracking to isolate particles and track them in consecutive frames; 7) data post-processing to obtain the relevant flow parameters. Obviously, flow images are acquired at a certain space–time resolution, depending on the characteristic time and length scales of the investigated phenomena, the details for each apparatus are provided in the corresponding subsection.

Pre-processing includes the sequence of operations carried out to improve the quality of acquired images for the subsequent core of the processing phase. Basically, the procedure implies the background removal as well as the removal of parts of the image which are not significant for the flow analysis as for instance regions close to the boundaries. In fact, the glares due to the interaction between the lighting system and the domain walls may affect the processing algorithm.

FT is a multi-frame algorithm based on the assumption of image light intensity conservation in space and time between two successive frames and in the neighborhood of the seeding particles; this assumption holds for small time intervals. The algorithm essentially considers measures of correlation windows between successive frames and evaluates displacements by considering the best correspondence (in terms of a defined matching measure) of selected interrogation windows between subsequent images. Sparse velocity vectors are then obtained by dividing the displacement by the time interval between two frames; FT then provides a Lagrangian description of the velocity field. These sparse data can be interpolated on a regular grid through a resampling procedure allowing for the reconstruction of the instantaneous and time-averaged Eulerian velocity fields as well. The advantage of having at the same time both the Lagrangian and the Eulerian description of the flow is evident; in addition, if compared to other tracking algorithms, FT is not constrained by low seeding density, so it provides accurate displacement vectors even when the number of tracer particles within each image is very large [22].

3. Data analysis

3.1 Traveling waves and eddies

As mentioned before, in these jet flows waves and eddies co-exist; to highlight the propagation of the traveling structures in the physical space, we consider both a measure based on Hovmöller diagrams and the theoretical phase speed of the Rossby wave.

As for the former, we map the time evolution of the stream function ψ as a function of φ at different radius, and in particular in correspondence of $r = r_{MS}$, i.e., the radius where the radial shear is a maximum. The diagrams may show linear features with negative or positive slopes, indicating westward/eastward propagating structures; the propagating structures could be waves or eddies, or both. To calculate the propagation velocity V_p in correspondence of a radius r we estimate the slope $\Delta\varphi/\Delta t$ of the contour lines in the azimuthal diagrams:

$$V_p = r\frac{\Delta\varphi}{\Delta t} \tag{1}$$

then the net speed of the propagating structures is evaluated by subtracting the mean zonal velocity from V_p:

$$V_{pn} = V_p - \langle V_z \rangle_\varphi \tag{2}$$

where $\langle V_z \rangle_\varphi$ is the mean zonal velocity averaged over a range of φ corresponding to the forced sector and time interval of ~300 s.

As for the theoretical speed, we have shown in [14] how to derive the dispersion relation of a linear Rossby wave in polar coordinates; here, we reported the final expression:

$$V_t = U - \beta\frac{R^2}{\alpha^2} \tag{3}$$

being R the radius of the device (in this case the radius of the circle inscribed in the square tank), U the average zonal velocity in correspondence of the chosen radius r, and α the coefficient of the Bessel Fourier decomposition, depending on the geometry of the system and on the width of the forced sector and the characteristic of the forcing.

In oceanography, one of the most popular methods used to detect coherent long-lived coherent structures, such as mesoscale eddies, is based on the estimation of the Okubo-Weiss parameter [24, 25]. This quantity describes the relative dominance of deformation with respect to rotation of the flow and it is defined as:

$$OW = s_n - s_s - \omega^2 \tag{4}$$

where $s_n = u_x - v_y$ and $s_s = v_x + u_y$ are the normal and shear components of strain, respectively, $\omega = v_x - u_y$ is vorticity. The subscripts $()_x$ and $()_y$ indicate partial differentiation of the horizontal velocities (u and v) in the x and y directions, respectively. In order to distinguish regions characterized by different topology within the flow domain, one has first to fix a positive threshold OW_0 of the Okubo-Weiss parameter. Then, according to it, the domain can be divided into zones corresponding to vortex cores ($OW < -OW_0$), organized structures surrounding vortex cores ($OW > OW_0$), and the background field ($|OW| \leq OW_0$). A value typically assumed for the threshold is $OW_0 = 0.2\sigma_{OW}$, where σ_{OW} is the standard deviation of OW parameter [26, 27].

3.2 Finite-time Lyapunov exponents and Lagrangian coherent structures

Finite-Time Lyapunov Exponents (FTLE) represents a powerful tool suitable to track coherent structures and to unveil their connections to energetic and mixing processes, in fact, it has been used extensively in different contexts, including biological and geophysical flows [28, 29]. Basically, the FTLE measure the maximum linearized growth rate of the distance among initially adjacent particles tracked over a finite integration time. In brief, the computation of FTLE follows from the definition of the flow map $\Phi(x)_t^{t+T^*}$ over a finite time interval T^*:

$$\Phi(x)_t^{t+T^*} : \quad x(t) \rightarrow x(t + T^*) \tag{5}$$

mapping a material point x(t) at time t to its position at t + T^* along its trajectory. After linearization, the amount of stretching about a trajectory is defined in terms of the Cauchy-Green deformation tensor by the matrix:

$$\Delta = \left(\frac{d\Phi(x)_t^{t+T^*}}{dt} \right)^2 \tag{6}$$

Since the maximum stretching occurs when the initial separation is aligned with the maximum eigenvalue of Δ, the FTLE is defined as:

$$\sigma(x, t, T^*) = \frac{1}{|T^*|} \ln \sqrt{\lambda_{max}} \tag{7}$$

Where λ_{max} is the maximum eigenvalue of Δ and $\sqrt{\lambda_{max}}$ corresponds to the maximum stretching factor. In particular, if a positive time interval is considered, the FTLEs measure separation forward in time, thus identifying repelling structures. On the contrary, if negative time intervals are considered, FTLEs measure separation backward in time, thus highlighting attracting structures [28, 30].

In addition, Lagrangian Coherent Structures (LCS) can be inferred from FTLE, [31]. LCS analysis represents a very powerful tool in cardiovascular fluid dynamics [32]; allowing for the identification of stagnant fluid areas, which are associated with an increased risk of thrombus as well as with blood cell damage. In addition, it helps to discern the regions directly affected by the vortices within the fluid domain and, possibly, their, modifications related to pathologies. FTLE investigation was successfully applied to the analysis of data sets obtained from both numerical simulations [27] and in vitro study [33] of a mechanical heart valve, as well as for the in vitro investigation of coherent structures educed from two-dimensional velocity fields in a LV model [21]. Recently, FTLE is also being used in the analysis of data sets collected in vivo [34, 35] and have been recognized as one of the main methods for the analysis of Lagrangian transport in blood flows [36, 37].

4. Results and discussion

4.1 Rotating flows affected by a β-effect

Before running each experiment, the fluid surface is seeded with styrene particles (mean size d_m = 50 μm) acting as passive tracers and the fluid surface is lighted with two lateral lamps. The rotation rate of the table is then raised up to the chosen value and, once the solid body rotation is established, the forcing is switched on, and flow images are acquired at 20fps by a high-resolution video camera (1023x1240 pixels) co-rotating with the system, perpendicular to the tank and with the optical axis parallel to the rotation axis. FT allows to reconstruct particle trajectories i.e., to provide a description of the flow in a Lagrangian framework; once the instantaneous sparse velocity vectors have been detected, they are interpolated onto a regular grid. In this case, it was convenient to choose a polar coordinate (r, φ) system with the pole corresponding to the center of the tank: the azimuthal direction φ identifies points with the same fluid depth (the so-called zonal direction) and at constant radius r. Sparse data have then been rearranged on a polar grid with 120 radii and 60 circles using a standard cubic spline interpolation procedure. The non-dimensional parameters of importance in our model are: the aspect ratio i.e., the ratio between the horizontal and vertical dimension of the flow domain $H_0/L < <1$ (shallow fluid); the Ekman number Ek $= \nu / \Omega H_0^2$ is order $O(10^{-4})$, ν is the kinematic viscosity; the Rossby number Ro $= U/2\Omega L$ order $O(10^{-3})$, U is the velocity scale; the Reynolds number Re $= Ul_\nu/\nu$ is order $O(10^2)$, l_ν is the characteristic length scale of the eddies. We summarize here some of the main results obtained in the characterization of eastward/westward flows, hereafter indicated as EW and WW case.

4.1.1 Waves and eddies propagation

In **Figures 1** and **2** we plot the instantaneous and time-averaged flow fields obtained in one run (I = 4A) of the experiments WW and EW; the plots are shown hereafter refer to experiments performed using the same forcing amount. **Figure 1** clearly shows a meandering jet squeezed between westward propagating eddies in the instantaneous flow field; on the contrary, the averaged field reveals strong alternating zonal jets and no eddies. These experimental features resemble ocean observations that highlight numerous westward propagating eddies on short time scales [12]. At the difference, the eastward jet is not associated with eddy shedding and traveling structures and the instantaneous and averaged flow appear to be rather similar (**Figure 2**).

To characterize the traveling structure observed in the WW case, we map the velocity stream function ψ as a function of time t and longitude φ in correspondence of the radius of maximum radial shear, r_{MS} (**Figure 3**). In fact, in [14] we were able to demonstrate that the best match between the theoretical and experimental estimation of the speed of propagating structures is found in the correspondence of this radius. To emphasize this aspect, we plot in **Figure 4**, from left to right: the

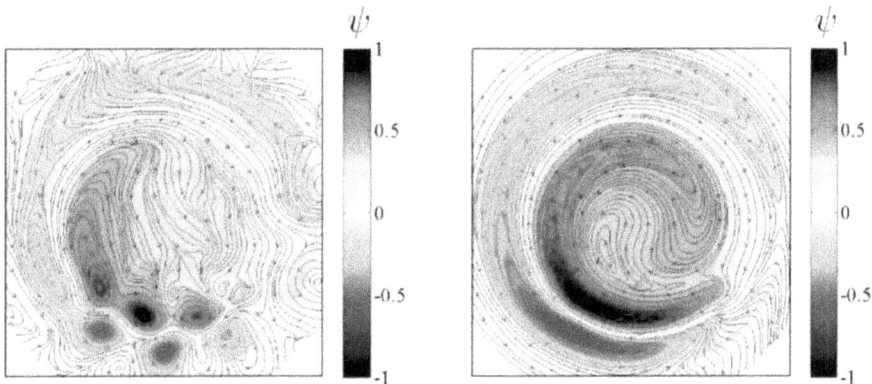

Figure 1.
Instantaneous (left) and time mean (right) normalized stream function superimposed on the streamlines for a WW flow.

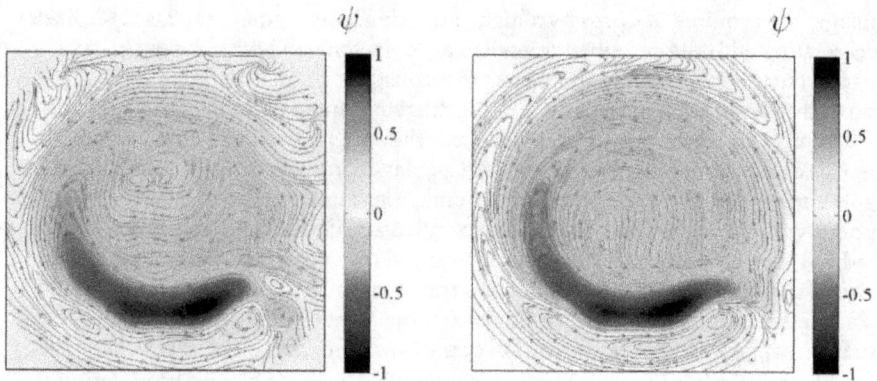

Figure 2.
Instantaneous (left) and time mean (right) normalized stream function superimposed on the streamlines for a EW flow.

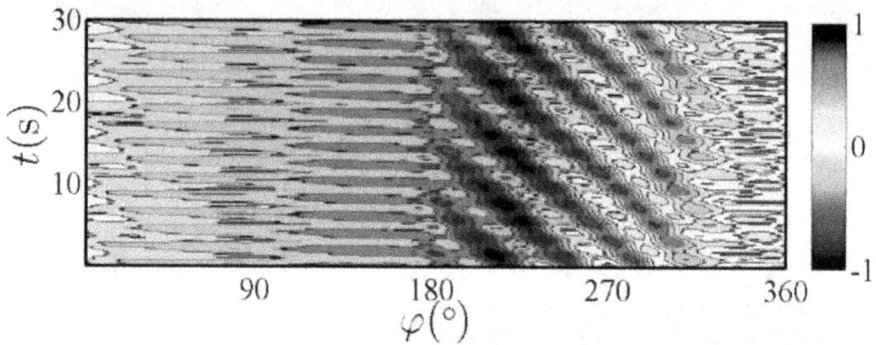

Figure 3.
Azimuthal Hovmöller diagram of the stream function ψ with the radius of the maximum radial shear chosen as the reference radius, WW flow.

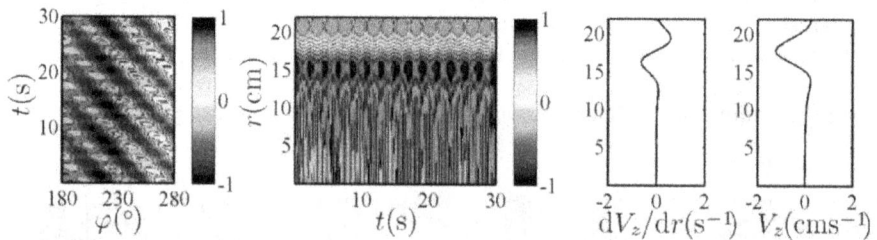

Figure 4.
From left to right: Azimuthal Hovmöller diagram of the stream function ψ with the radius of the maximum radial shear chosen as the reference radius; radial Hovmöller diagram of ψ averaged azimuthally in the range forced sector; radial shear profiles of the azimuthal velocity V_z averaged azimuthally in the same sector and in time.

azimuthal (φ-t) Hovmöller diagram of ψ, the radial t-r Hovmöller diagram of ψ, the radial profiles of the mean radial shear $d\langle \overline{V_Z} \rangle_\varphi / dr$ and the mean azimuthal velocity $\langle \overline{V_Z} \rangle_\varphi$. The profiles show a maximum of the mean zonal velocity in correspondence of the radius r_C corresponding to the jet centerline while the maximum of the mean shear is at a radius $r_{MS} > r_C$. We recall that the definition of the jet boundaries and of its time evolution is crucial in the definition of the barriers to meridional transport [5, 13].

$$\zeta(s^{-1}) \qquad\qquad Q_{OW}(s^{-2})$$

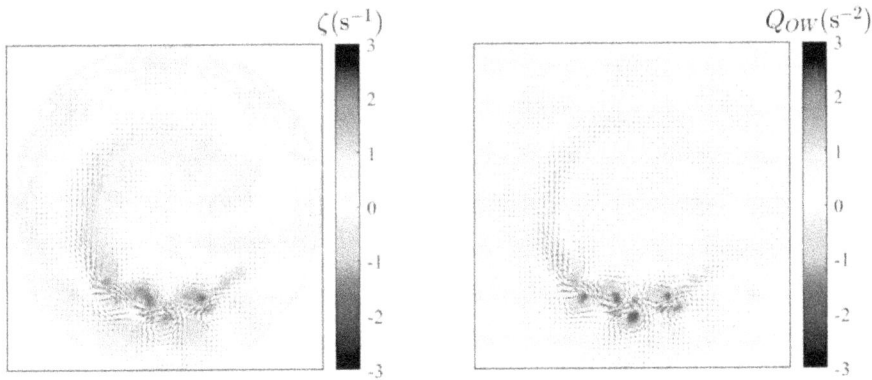

Figure 5.
Instantaneous fields of vorticity field ζ (left) and Okubo–Weiss parameter Q_{ow} (right); velocity field superimposed (blue arrows) for a WW flow.

As discussed in Section 3.1, by measuring the slope of the lines of the same color, we were able to estimate experimentally the speed of the propagating structures relative to the zonal flow with Eq. (2); we then calculate the theoretical speed using Eq. (3) and compare the obtained values. The comparison shows that the relative error, i.e. the ratio between the measured and the expected speed, is minimum in correspondence of r_{MS} ($O(10^{-2})$).

In order to compare our method to evaluate the eddies propagation speed with a method widely used in the applications we also applied an OW-based method to our experimental data sets. At first, we evaluated OW parameter through Eq. (4) at each time instant. Then, using a threshold of $OW_0 = 0.5\sigma_{OW}$ we identified the vortex cores and their surrounding area. We refined the detection by combining the OW parameter with the physical properties of the flow field (high vorticity areas, velocity vectors) and by applying geometrical constraints. An example is provided in **Figure 5** in which we show a snapshot of the vorticity field ζ (left) and of the Okubo–Weiss parameter Q_{OW} (right) superimposed on the corresponding velocity field. In the vorticity map, regions of dark blue (dark red) identify strong anti-cyclonic (cyclonic) circulation. In the Q_{ow} field, dark blue identifies regions where vorticity is much stronger than strain (i.e., eddy cores), and dark red where strain is much greater than vorticity.

Finally, once identified the coherent vortices, we detected the center of each structure and tracked them in the considered time interval. We found values of the propagating speed close to the ones found through the Hovmöller diagrams. We conclude that waves and propagating eddies coexist in the zonal pattern and confirm their duality nature [14]. The application of the same procedure of analysis overall the EW experiments is actually in progress [38].

4.1.2 Characteristic scales and flow regime

The estimation of flow characteristic length scales is crucial to identify the flow regime in rotating turbulent flows with a β-effect. To this aim, as discussed in Section 2.1, we calculate the Rhines number, k_{RH}, the transitional wave number, k_β, and their ratio R_β. The results for a set of WW and EW experiments are reported in **Table 1**.

According to the classification provided in [8] we conclude that all our experiments reproduced flows in a transitional regime.

Run	$k_{Rh}(\mathrm{cm}^{-1})$	$k_\beta\ (\mathrm{cm}^{-1})$	R_β
WW$_1$	1.41	2.20	1.55
WW$_2$	1.09	1.75	1.60
WW$_3$	0.94	1.16	1.70
EW$_1$	1.19	1.90	1.59
EW$_2$	0.85	1.45	1.70
EW$_3$	0.74	1.29	1.75

Table 1.
Characteristic scales and zonostrophy index estimated from experimental data.

4.2 Flow patterns in the left ventricle downstream of prosthetic valves

To perform flow measurements in the LV, the vertical symmetry plane aligned with the mitral and aortic valve axes is illuminated by a 12 W, infrared laser. The working fluid inside the ventricle (distilled water) is seeded with neutrally buoyant particles ($d_m \sim 30$ μm). A high-speed digital camera (250 frames/s, 1280 × 1024 pixel resolution) is triggered by the motor to frame the time evolution of the phenomenon for N cardiac cycles. The acquired images are processed by means of a FT algorithm and velocity fields on a regular grid 51 × 51 are obtained for the considered time interval. Two-dimensional Eulerian velocity data were then phase averaged over N = 50 cycles. Here, we discuss two groups of experiments performed considering a period T = 6 s and stroke volumes SV_1 = 64 ml, SV_2 = 80 ml. We briefly recall that during the cardiac cycle the flow rate change according to the considered law [21]: the fluid enters the LV through the mitral valve during the diastole (0.00 T – 0.75 T) and is ejected out through the aortic valve during the systole (0.75 T – 1.00 T). Two peaks separated by the diastasis characterize the diastole phase: the first is called E-wave and corresponds to the dilation of the ventricle, and the second, called A-wave, is due to the contraction of the left atrium.

For the dynamic similarity, we consider the Reynolds number $Re = UD/v$ and the Womersley number $Wo = \sqrt{D^2/Tv}$; respectively equal to Re_1 = 8322; Re_2 = 10403; Wo = 22.8; in both cases within the physiological range. Here, D is the maximum diameter of the ventricle, U the peak velocity through the mitral orifice, v the kinematic viscosity of the working fluid i.e., distilled water.

We use the public domain code NEWMAN [39] to compute the FTLEs from the planar velocity dataset above described, for the details see [40]. We remark that FTLE fields are computed from 2D measurements even if it is well known that the observed phenomenon is 3D; indeed, as the measurement plane is a plane of symmetry the assumption of two-dimensionality is quite acceptable.

Figure 6 shows backward FTLE at the end of the E-wave for the three simulated conditions. Backward FTLE ridges correspond to the front of the diastolic jet, sharply separating the fluid which just entered the ventricle from the receiving fluid.

The analysis of the FTLE patterns throughout the cardiac cycle (not shown here) highlights how in the physiological configuration the observed coherent structures appear to be optimal for the systolic function. Indeed, the modifications in the transmitral flow due to the presence of a prosthetic valve deeply impact on the interaction between the coherent structures generated during the first phase of the diastole and the incoming jet during the second diastolic phase. We observed that

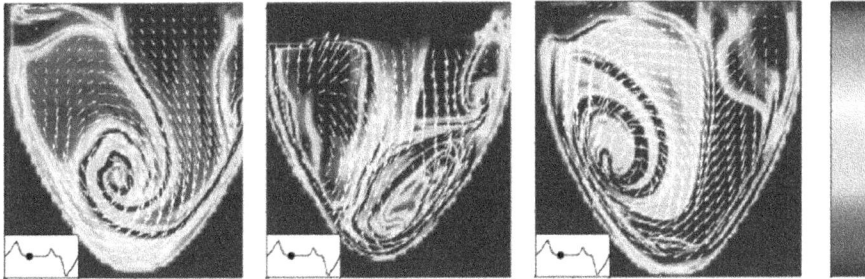

Figure 6.
Velocity fields and backward FTLE at the end of E-wave (the small inset shows the current time in the cardiac cycle as a black dot): Left: Physiological configuration, center: Monoleaflet valve, and right: Bileaflet valve.

while the flow generated by a bileaflet valve preserves most of the beneficial features of the top hat inflow, downstream of a monoleafleat one the strong jet forming at the end of the diastole prevents the permanence of large coherent structures within the LV (**Figure 7**).

In order to complete the FTLE analysis, we reconstruct the trajectories of a number $(O(10^4))$ of synthetic fluid particles entering the ventricle through the mitral orifice during the LV filling by numerically integrating the experimental velocity fields; for each run, synthetic particles were released during each time step of the diastolic waves from the mitral orifice section and were subsequently tracked during the cardiac cycle. The aim was to further clarify the role of LCS by overlapping the

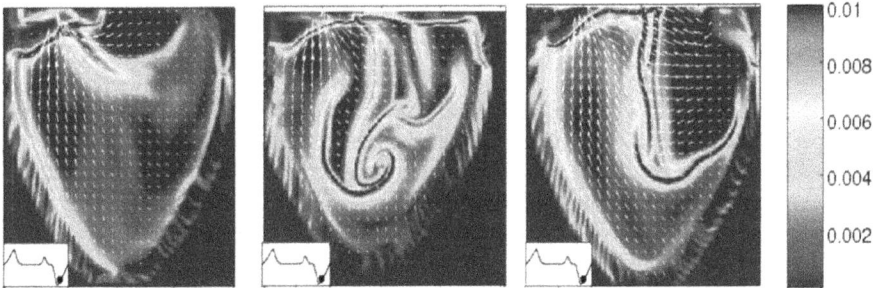

Figure 7.
Same as above in correspondence of the systolic peak.

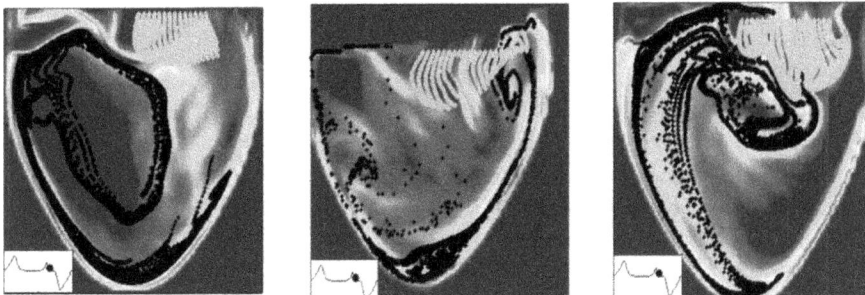

Figure 8.
Syntetic particles overlapped on FTLE maps at the end of the a wave.

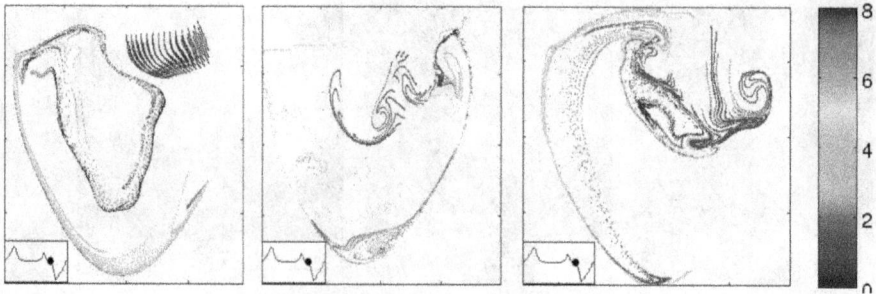

Figure 9.
Synthetic particles entered through the mitral orifice during diastolic waves colored according to the shear stress cumulated until the end of the A-wave (the color bar values correspond to the non-dimensional maximum shear stress).

particle positions on the FTLE maps, and to verify if and how LCS may act as pseudo-barriers for transport and mixing. An example is reported in **Figure 8**.

We finally compute the shear stress experienced by the particles along their trajectories in order to emphasize the differences among the simulated conditions and to clarify the possible implications on the hemodynamics. Results corresponding to the end of the A wave are shown in **Figure 9**.

The plots show that, in case (a) the stress magnitudes induced by the smoother flow pattern are lower than values measured in case (b) and (c). In fact, while in physiological conditions particles characterized by the highest shear are washed out by the systolic wave, in presence of prosthetic valves they tend to be advected towards regions of the LV not affected by the systolic ejection (see **Figure 3**).

5. Conclusions

In this work, we review a set of methodologies suitable for the characterization of time-periodic complex flows; in particular, here, the focus is on rotating flows affected by a β-effect and blood flow in the left ventricle. The interest in deepening these contexts depends on their importance from both an applicative and a methodological point of view. Indeed, we consider almost 2D and time-periodic flows in which, due to chaotic advection, the Lagrangian motion of passive particles can be very complex even in regular, i.e., non-turbulent, Eulerian flow fields. We believe that the obtained results show, on one hand, that the designed experimental models prove suitable to reproduce the investigated phenomena, and on the other hand confirm that the proposed methodologies represent valid and powerful tools for identifying and characterizing the main flow patterns in their space–time evolution.

Acknowledgements

The authors would like to thank the Sapienza University of Rome (Research program SAPIEXCELLENCE SPC: 2021-1136-1451-173491), the European Union's Horizon 2020 research and innovation program (Marie Sklodowska–Curie Grant Agreement No. 797012) and the Italian Ministry of Research (project PRIN 2017 A889FP).

Conflict of interest

The authors declare no conflict of interest.

Author details

Stefania Espa[1*], Maria Grazia Badas[2] and Simon Cabanes[1]

1 Sapienza University of Rome, Rome, Italy

2 University of Cagliari, Cagliari, Italy

*Address all correspondence to: stefania.espa@uniroma1.it

IntechOpen

References

[1] Crisanti A, Falcioni M, Paladin G, Vulpiani A. Lagrangian chaos:Transport, mixing and diffusion in fluids. La Rivista del Nuovo Cimento. 1991;**14**: 1-80. DOI: 10.1007/BF02811193

[2] Rhines PB. Waves and turbulence on a beta-plane. Journal of Fluid Mechanics. 1975;**69**:417-443. 69(3), 417-443. DOI: 10.1017/S0022 112075001504

[3] Cho JYK, Polvani LM. The emergence of jets and vortices in freely evolving, shallow-water turbulence on sphere. Physics of Fluids. 1996;**8**: 1531-1552. DOI: 10.1063/1.868929

[4] Maximenko NA, Bang B, Sasaki H. Observational evidence of alternating zonal jets in the World Ocean. Geophysical Research Letters. 2005;**32**: L12607. DOI: 10.1029/2005GL022728

[5] Galperin B, Read PL (Eds.). Zonal Jets: Phenomenology, Genesis, and Physics. Cambridge University Press; 2019. p. 550. DOI: 10.1017/ 9781107358225

[6] Galperin B, Sukoriansky S, Dikovskaya N. Geophysical flows with anisotropic turbulence and dispersive waves: Flows with a β-effect. Ocean Dynamics. 2010;**60**:427-441. DOI: 10.1007/s10236-010-0278-2

[7] Kraichnan RH. Inertial ranges in two-dimensional turbulence. Physics of Fluids. 1967;**10**:1417. DOI: 10.1063/ 1.1762301

[8] Sukoriansky S, Dikovskaya N, Galperin B. On the arrest of the inverse energy cascade and the Rhines scale. Journal of the Atmospheric Sciences. 2007;**64**:3312-3327. DOI: 10.1175/ JAS4013.1

[9] Espa S, Di Nitto G, Cenedese A. The emergence of zonal jets in forced rotating shallow water turbulence: A laboratory study. Europhysics Letters. 2010;**92**:34006. DOI: 10.1209/ 0295-5075/92/34006

[10] Di Nitto G, Espa S, Cenedese A. Simulating zonation in geophysical flows by laboratory experiments. Physics of Fluids. 2013;**25**:086602. DOI: 10.1063/1.4817540

[11] Derzho OG, Afanasyev YD. Rotating dipolar gyres on a β-plane. Physics of Fluids. 2008;**20**:036603. DOI: 10.1063/ 1.2890083

[12] Galperin B, Hoemann J, Espa S, Di Nitto G. Anisotropic turbulence and Rossby waves in an easterly jet: An experimental study. Geophysical Research Letters. 2014;**41**:6237-6243. DOI: 10.1002/2014gl060767

[13] Galperin B, Hoemann J, Espa S, Di Nitto G, Lacorata G. Anisotropic macroturbulence and diffusion associated with a westward zonal jet: From laboratory to planetary atmospheres and oceans. Physical Review E. 2016;**94**:063102. DOI: 10.1103/PhysRevE.94.063102

[14] Espa S, Cabanes S, King GP, Di Nitto G, Galperin B. Eddy-wave duality in a rotating flow. Physics of Fluids. 2020;**32**:076604. DOI: 10.1063/ 5.0006206

[15] Cabanes S, Espa S, Galperin B, Young RMB, Read PL. Revealing the intensity of turbulent energy transfer in planetary atmospheres. Journal of Geophysical Research. 2020;**47**:2316. DOI: 10.1029/2020GL088685

[16] Pedrizzetti G, Domenichini F. Nature optimizes the swirling flow in the human left ventricle. Physical Review Letters. 2005;**95**:108101. DOI: 10.1103/PhysRevLett.95.108101

[17] Yoganathan AP, He Z, Casey Jones S. Fluid mechanics of heart valves. Annual Review of Biomedical Engineering. 2004;**6**:331-362. DOI: 10.1146/annurev.bioeng.6. 040803.140111

[18] Sotiropoulos F, Le TB, Gilmanov A. Fluid mechanics of heart valves and their replacements. Annual Review of Fluid Mechanics. 2016;**48**:259-283. DOI: 10.1146/annurev-fluid-122414-034314

[19] Querzoli G, Cenedese A, Fortini S. Effect of the prosthetic mitral valve on vortex dynamics and turbulence of the left ventricular flow. Physics of Fluids. 2010;**22**:041901. DOI: 10.1063/ 1.3371720

[20] Fortini S, Querzoli G, Espa S, Cenedese A. Three-dimensional structure of the flow inside the left ventricle of the human heart. Experiments in Fluids. 2013;**54**:1-9. DOI: 10.1007/s00348-013-1609-0

[21] Espa S, Badas MG, Fortini S, Querzoli G, Cenedese A. A Lagrangian investigation of the flow inside the left ventricle. European Journal of Mechanics - B/Fluids. 2012;**35**:9-19. DOI: 10.1016/j.euromechflu.2012.01.015

[22] Moroni M, Cenedese A. Comparison among feature tracking and more consolidated velocimetry image analysis techniques in a fully developed turbulent channel flow. Measurement Science and Technology. 2005;**16**:2307. DOI: b10.1088/0957-0233/16/11/025

[23] Funiciello F, Moroni M, Piromallo C, Faccenna C, Cenedese A, Bui HA. Mapping mantle flow during retreating subduction: Laboratory models analyzed by feature tracking. Journal of Geophysical Research. 2006; **111**:B03402-B03412. DOI: 10.1029/ 2005GL025390

[24] Okubo A. Horizontal dispersion of floatable particles in the vicinity of velocity singularities such as convergences. Deep Sea Research. 1970; **17**:445-454. DOI: 10.1016/0011-7471 (70)90059-8

[25] Weiss J. The dynamics of enstrophy transfer in two-dimensional hydrodynamics. Physica D. 1991;**48**: 273-294. DOI: 10.1016/0167-2789(91) 90088-Q

[26] Elhmaïdi D, Provenzale A, Babiano A. Elementary topology of two dimensional turbulence from a Lagrangian viewpoint and single-particle dispersion. Journal of Fluid Mechanics. 1993;**257**:533-558. DOI: 10.1017/S0022112093003192

[27] Pasquero C, Provenzale A, Babiano A. Parameterization of dispersion in two-dimensional turbulence. Journal of Fluid Mechanics. 2001;**439**:279-303. DOI: 10.1017/ S0022112001004499

[28] Shadden SC, Lekien F, Marsden JE. Definition and properties of Lagrangian coherent structures from finite-time Lyapunov exponents in two-dimensional aperiodic flows. Physica D: Nonlinear Phenomena. 2005;**212**: 271-304. DOI: 10.1016/j.physd.2005. 10.007

[29] Beron-Vera FJ, Olascoaga M, Goni G. Oceanic mesoscale eddies as revealed by Lagrangian coherent structures. Geophysical Research Letters. 2008;**35**:12. DOI: 10.1029/ 2008GL033957

[30] Haller G. Distinguished material surfaces and coherent structures in three-dimensional fluid flows. Physica D. 2001;**149**:248-277. DOI: 10.1016/S0167-2789(00)00199-8

[31] Haller G. Lagrangian coherent structures. Annual Review of Fluid Mechanics. 2015;**47**:137-162. DOI: 10.1146/annurev-fluid-010313-141322

[32] Shadden SC, Taylor CA. Characterization of coherent structures in the cardiovascular system. Annals of Biomedical Engineering. 2008;**36**(7): 1152-1162. DOI: 10.1007/s10439-008-9502-3

[33] Miron P, Vétel J, Garon A. On the use of the finite-time Lyapunov exponent to reveal complex flow physics in the wake of a mechanical valve. Experiments in Fluids. 2014;**55**: 1-15. DOI: 10.1007/s00348-014-1814-5

[34] Hendabadi S, Bermejo J, Benito Yotti R, Fernández-Avilés F, del Álamo JC, Shadden SC. Topology of blood transport in the human left ventricle by novel processing of Doppler echocardiography. Annals of Biomedical Engineering. 2013;**41**:2603-2616. DOI: 10.1007/s10439-013-0853-z

[35] Bermejo J, Benito Y, Alhama M, Yotti R, Martínez-Legazpi P, Del Villar CP, et al. Intraventricular vortex properties in nonischemic dilated cardiomyopathy. AJP - Heart and Circulatory Physiology. 2014;**306**: H718-H729. DOI: 10.1152/ajpheart. 00697.2013

[36] Badas MG, Domenichini F, Querzoli G. Quantification of the blood mixing in the left ventricle using finite time Lyapunov exponents. Meccanica. 2017;**52**(3):529-544. DOI: 10.1007/ s11012-016-0364-8

[37] Di Labbio G, Vétel J, Kadem L. Material transport in the left ventricle with aortic valve regurgitation. Physical Review Fluids. 2021;**6**:059901. DOI: 10.1103/PhysRevFluids.6.059901

[38] Espa S, Cabanes S. Eddies and waves in a rotating flow: An experimental study. In: Proceedings of the 39th IAHR World Congress 19–24 June 2022, Granada, Spain

[39] Du Toit PC, Marsden JE. Horseshoes in hurricanes. Journal of Fixed Point Theory and Applications 2010;7(2):351-384. DOI: 10.1007/s11784-010-0028-6.

[40] Badas MG, Espa S, Fortini S, Querzoli G. 3D finite time Lyapunov exponents in a left ventricle laboratory model. EPJ Web of Conferences. 2015; **926**:02004. DOI: 10.1051/epjconf/ 20159202004

Chapter 6

Relaxation Dynamics of Point Vortices

Ken Sawada and Takashi Suzuki

Abstract

We study a model describing relaxation dynamics of point vortices, from quasi-stationary state to the stationary state. It takes the form of a mean field equation of Brownian point vortices derived from Chavanis, and is formulated by our previous work as a limit equation of the patch model studied by Robert-Someria. This model is subject to the micro-canonical statistic laws; conservation of energy, that of mass, and increasing of the entropy. We study the existence and nonexistence of the global-in-time solution. It is known that this profile is controlled by a bound of the negative inverse temperature. Here we prove a rigorous result for radially symmetric case. Hence E/M^2 large and small imply the global-in-time and blowup in finite time of the solution, respectively. Where E and M denote the total energy and the total mass, respectively.

Keywords: point vortex, quasi-equilibrium, relaxation dynamics

1. Introduction

Our purpose is to study the system

$$\omega_t + \nabla \cdot \omega \nabla^{\perp}\psi = \nabla \cdot (\nabla\omega + \beta\omega\nabla\psi) \quad \text{in } \Omega \times (0,T),$$
$$\frac{\partial\omega}{\partial\nu} + \beta\omega\frac{\partial\psi}{\partial\nu}\bigg|_{\partial\Omega} = 0, \quad \omega|_{t=0} = \omega_0(x) \tag{1}$$

with

$$-\Delta\psi = \omega \text{ in } \Omega, \quad \psi|_{\partial\Omega} = 0, \quad \beta = -\frac{\int_{\Omega}\nabla\omega \cdot \nabla\psi}{\int_{\Omega}\omega|\nabla\psi|^2}, \tag{2}$$

where $\Omega \subset \mathbf{R}^2$ is a bounded domain with smooth boundary $\partial\Omega$, ν is the outer unit normal vector on $\partial\Omega$, and

$$\nabla = \begin{pmatrix} \dfrac{\partial}{\partial x_1} \\[2mm] \dfrac{\partial}{\partial x_2} \end{pmatrix}, \quad \nabla^{\perp} = \begin{pmatrix} \dfrac{\partial}{\partial x_2} \\[2mm] -\dfrac{\partial}{\partial x_1} \end{pmatrix}, \quad x = (x_1, x_2). \tag{3}$$

The unknown $\omega = \omega(x,t) \in \mathbf{R}$ stands for a mean field limit of many point vortices,

IntechOpen

$$\omega(x,t)dx = \sum_{i=1}^{N} \alpha_i \delta_{x_i(t)}(dx). \tag{4}$$

It was derived, first, for Brownian point vortices by [1, 2], with $\beta = \beta(t)$ standing for the inverse temperature. Then, [3, 4] reached it by the Lynden-Bell theory [5] of relaxation dynamics, that is, as a model describing the movement of the mean field of many point vortices, from quasi-stationary state to the stationary state. This model is consistent to the Onsager theory [6–12] on stationary states and also the patch model proposed by [13, 14], that is,

$$\omega(x,t) = \sum_{i=1}^{N_p} \sigma_i 1_{\Omega_i(t)}(x), \tag{5}$$

where N_p, σ_i, and $\Omega_i(t)$ denote the number of patches, the vorticity of the i-th patch, and the domain of the i-th patch, respectively [15–17].
This chapter is concerned on the one-sided case of

$$\omega_0 = \omega_0(x) > 0. \tag{6}$$

If this initial value is smooth, there is a unique classical solution to (1)–(4) local in time, denoted by $\omega = \omega(x,t)$, with the maximal existence time $T = T_{max} \in (0, +\infty]$. More precisely, the strong maximum principle to (1) guaranttes

$$\omega = \omega(x,t) > 0 \quad \text{on } \overline{\Omega} \times [0, T). \tag{7}$$

Then, the Hopf lemma to the Poisson equation in (2) ensures

$$\left.\frac{\partial \psi}{\partial \nu}\right|_{\partial\Omega} < 0, \tag{8}$$

and hence the well-definedness of

$$-\beta = \frac{\int_\Omega \nabla\omega \cdot \nabla\psi}{\int_\Omega \omega |\nabla\psi|^2}. \tag{9}$$

We confirm that system (1)–(3) satisfies the requirements of isolated system of thermodynamics. First, the mass conservation is derived from (1) as

$$\frac{d}{dt}\int_\Omega \omega = 0, \tag{10}$$

because

$$\nu \cdot \nabla^\perp \psi|_{\partial\Omega} = 0 \tag{11}$$

holds by (2). Second, the energy conservation follows as

$$\frac{1}{2}\frac{d}{dt}\|\nabla\psi\|_2^2 = (\nabla\psi, \nabla\psi_t) = (\omega_t, \psi)$$

$$= (\omega\nabla^\perp\psi, \nabla\psi) - (\nabla\omega + \beta\omega\nabla\psi, \nabla\psi) \tag{12}$$

$$= -(\nabla\omega, \nabla\psi) - \beta \int_\Omega \omega|\nabla\psi|^2 = 0$$

by (1) and (2), because

$$\nabla^\perp \psi \cdot \nabla \psi = 0, \tag{13}$$

where $(\ ,\ )$ denotes the L^2 inner product. Third, the entropy increasing is achieved, writing (1) as

$$\omega_t = \nabla \cdot \omega(-\nabla^\perp \psi + \nabla(\log \omega + \beta \psi)), \quad \frac{\partial}{\partial \nu}(\log \omega + \beta \psi)\Big|_{\partial \Omega} = 0. \tag{14}$$

In fact, it then follows that

$$\int_\Omega \omega_t(\log \omega + \beta \psi) = \int_\Omega \omega \nabla^\perp \psi \cdot \nabla(\log \psi + \beta \psi) - \omega |\nabla(\log \omega + \beta \psi)|^2 \, dx \tag{15}$$

with

$$\int_\Omega \omega \nabla^\perp \psi \cdot \nabla(\log \omega + \beta \psi) = \int_\Omega \nabla \omega \cdot \nabla^\perp \psi$$

$$= \int_{\partial \Omega} \omega \nu \cdot \nabla^\perp \psi - \int_\Omega \omega \nabla \cdot (\nabla^\perp \psi) = 0 \tag{16}$$

from (11) and

$$\nabla^\perp \cdot \nabla = \nabla \cdot \nabla^\perp = 0. \tag{17}$$

Since

$$\int_\Omega \omega_t \log \omega = \frac{d}{dt}\int_\Omega \omega(\log \omega - 1), \quad \int_\Omega \omega_t \psi = \frac{1}{2}\frac{d}{dt}\|\nabla \psi\|_2^2 = 0, \tag{18}$$

We thus end up with the mass conservation

$$M = \int_\Omega \omega, \tag{19}$$

the energy conservation

$$E = \|\nabla \psi\|_2^2 = (\psi, \omega), \tag{20}$$

and the entropy increasing

$$\frac{d}{dt}\int_\Omega \omega(\log \omega - 1) = -\int_\Omega \omega |\nabla(\log \omega + \beta \psi)|^2 \le 0. \tag{21}$$

Henceforth, $C > 0$ stands for a generic constant. In the previous work [4] we studied radially symmetric solutions and obtained a criterion for the existence of the solution global in time. Here, we refine the result as follows, where $B(0,1)$ denotes the unit ball.

Theorem 1 *Let*

$$\Omega = B(0,1), \quad \omega_0 = \omega_0(r), \quad \omega_{0r} < 0, \quad 0 < r = |x| \le 1. \tag{22}$$

Then there is $C_0 > 0$ such that

$$C_0 \|\omega_0\|_2^3 \le E\underline{\omega} \Rightarrow T = +\infty, \quad \|\omega(\cdot, t)\|_\infty \le C, \quad t \ge 0, \tag{23}$$

where

$$\underline{\omega} = \min_{\overline{\Omega}} \omega_0 > 0. \tag{24}$$

Theorem 2 *Under the assumption of (22) there is $\delta_0 > 0$ such that*

$$\frac{E}{M^2} < \delta_0 \Rightarrow T < +\infty. \tag{25}$$

Remark 1 *Since*

$$\|\omega_0\|_2^3 = \left(\int_\Omega \omega_0^2\right)^{3/2} \ge \left(\underline{\omega}^{2/3} \int_\Omega \omega_0^{4/3}\right)^{3/2}$$

$$= \underline{\omega} \left(\int_\Omega \omega_0^{4/3}\right)^{3/2} \ge \underline{\omega} |\Omega|^{-1/2} \left(\int_\Omega \omega_0\right)^2 = \underline{\omega} |\Omega|^{-1/2} M^2 \tag{26}$$

the assumption (23) implies

$$\frac{E}{M^2} \ge C_0 |\Omega|^{-1/2}. \tag{27}$$

Therefore, roughly, the conditions $E/M^2 \gg 1$ and $E/M^2 \ll 1$ imply $T = +\infty$ and $T < +\infty$, respectively.

Remark 2 *The assumption (22) implies*

$$\beta = \beta(t) < 0, \quad 0 \le t < T, \tag{28}$$

and then we obtain Theorem 1. In other words, the conclusion of this theorem arises from (28), without (22).

Remark 3 *Since*

$$\frac{E}{M^2} = \frac{\int_\Omega |\nabla \psi|^2}{\left(\int_\Omega \omega\right)^2} \tag{29}$$

it holds that

$$\frac{E}{M^2} = \|\nabla c\|_2^2, \quad c = \frac{(-\Delta)^{-1} \omega_0}{\int_\Omega \omega_0} = \frac{\psi_0}{\int_{\partial\Omega} - \frac{\partial \psi_0}{\partial \nu}}, \tag{30}$$

where $\psi_0 = (-\Delta)^{-1} \omega_0$.

The system (1)–(4) thus obays a profile of the micro-canonical ensemble. In a system associated with the canonical ensemble, the inverse temperature β is a constant in (1) independent of t, with the third equality in (2) elimiated:

$$\omega_t + \nabla \cdot \omega \nabla^\perp \psi = \nabla \cdot (\nabla \omega + \beta \omega \nabla \psi), \quad \left.\frac{\partial \omega}{\partial \nu} + \beta \omega \frac{\partial \psi}{\partial \nu}\right|_{\partial\Omega} = 0, \quad \omega|_{t=0} = \omega_0(x) > 0$$

$$-\Delta \psi = \omega, \quad \psi|_{\partial\Omega} = 0. \tag{31}$$

Then there arise the mass conservation

$$\frac{d}{dt}\int_{\Omega}\omega = 0, \tag{32}$$

and the free energy decreasing

$$\frac{d}{dt}\int_{\Omega}\omega(\log\omega - 1) + \frac{\beta}{2}|\nabla\psi|^2 \, dx = -\int_{\Omega}\omega|\nabla(\log\omega + \beta\psi)|^2 \le 0. \tag{33}$$

The system (31) without vortex term,

$$\omega_t = \nabla\cdot(\nabla\omega + \beta\omega\nabla\psi), \quad \frac{\partial\omega}{\partial\nu} + \beta\omega\frac{\partial\psi}{\partial\nu}\Big|_{\partial\Omega} = 0, \quad \omega|_{t=0} = \omega_0(x) > 0$$

$$-\Delta\psi = \omega, \quad \psi|_{\partial\Omega} = 0. \tag{34}$$

is called the Smoluchowski-Poisson equation. This model is concerned on the thermodynamics of self-gravitating Brownian particles [18] and has been studied in the context of chemotaxis [19–23]. We have a blowup threshold to (34) as a consequence of the quantized blowup mechanism [19, 23]. The results on the existence of the bounded global-in-time solution [24–26] and blowup of the solution in finite time [27] are valid even to the case that β is a function of t as in $\beta = \beta(t)$. provided with the vortex term $\nabla\cdot\omega\nabla^{\perp}\psi$ on the right-hand side. We thus obtain the following theorems.

Theorem 3 *It holds that*

$$-\beta(t) \le \delta, \quad \|\omega_0\|_1 < 8\pi\delta^{-1} \Rightarrow T = +\infty, \quad \|\omega(\cdot,t)\|_\infty \le C \tag{35}$$

in (31), where $\delta > 0$ is arbitrary.

Theorem 4 *It holds that*

$$-\beta(t) \ge \delta, \quad \|\omega_0\|_1 > 8\pi\delta^{-1} \Rightarrow \exists\omega_0 > 0, \quad \|\omega_0\|_1 > 8\pi\delta^{-1} \quad \text{such that} \quad T < +\infty \tag{36}$$

in (31), where $\delta > 0$ is arbitrary.

Remark 4 *In the context of chemotaxis in biology, the boundary condition of ψ is required to be the form of Neumann zero. The Poisson equation in (34) is thus replaced by*

$$-\Delta\psi = \omega - \frac{1}{|\Omega|}\int_{\Omega}\omega, \quad \frac{\partial\psi}{\partial\nu}\Big|_{\partial\Omega} = 0 \tag{37}$$

or

$$-\Delta\psi + \psi = \omega, \quad \frac{\partial\psi}{\partial\nu}\Big|_{\partial\Omega} = 0 \tag{38}$$

by [28] and [29], respectively. In this case there arises the boundary blowup, which reduces the value 8π in Theorems 3–4 to 4π. The value 8π in Theorems 3–4, therefore, is a consequence of the exclusion of the boundary blowup [30]. This property is valid even for (37) or (38) of the Poisson part, if (22) is assumed.

Remark 5 *The requirement to ω_0 in Theorem 4 is the concentration at an interior point, which is not necessary in the case of (22). Hence Theorems 3 and 4 are refined as*

$$-\beta(t) \le \delta, \quad \|\omega_0\|_1 < 8\pi\delta^{-1} \Rightarrow T = +\infty, \quad \|\omega(\cdot,t)\|_\infty \le C \tag{39}$$

and

$$-\beta(t) \geq \delta, \quad \|\omega_0\|_1 > 8\pi\delta^{-1} \Rightarrow T < +\infty, \tag{40}$$

if (22) holds in (35). The main task for the proof of Theorems 1 and 2, therefore, is a control of $\beta = \beta(t)$ in (1).

This paper is composed of four sections and an appendix. Section 2 is devoted to the study on the stationary solutions, and Theorems 1 and 2 are proven in Sections 3 and 4, respectively. Then Theorem 4 is confirmed in Appendix.

2. Stationary states

First, we take the canonical system (31) with β independent of t. By (32) and (33), its stationary state is defined by

$$\log \omega + \beta\psi = \text{constant}, \quad \omega = \omega(x) > 0, \quad \int_\Omega \omega = M. \tag{41}$$

Then it holds that

$$\omega = \frac{Me^{-\beta\psi}}{\int_\Omega e^{-\beta\psi}} \tag{42}$$

and hence

$$-\Delta\psi = \frac{Me^{-\beta\psi}}{\int_\Omega e^{-\beta\psi}}, \quad \psi|_{\partial\Omega} = 0. \tag{43}$$

There arises an oredered structure arises in $\beta < 0$, as observed by [11], as a consequence of a quantized blowup mechanism [19, 20, 31]. In the microcanonical system (1) and (2), the value β in (43) has to be determined by E besides M.

Equality (21), however, still ensures (41) and hence (42) in the stationary state even for (1)–(3). Writing

$$v = -\beta\psi, \quad \mu = \frac{-\beta M}{\int_\Omega e^{-\beta\psi}}, \tag{44}$$

we obtain

$$-\Delta v = \mu e^v \text{ in } \Omega, \quad v|_{\partial\Omega} = 0, \quad \frac{E}{M^2} = \frac{\|\nabla v\|_2^2}{\left(\int_\Omega - \frac{\partial v}{\partial v}\right)^2} \tag{45}$$

by (30) and (43).

This system is the stationary state of (1) and (2) introduced by [4]. The first two equalities

$$-\Delta v = \mu e^v, \quad v|_{\partial\Omega} = 0 \tag{46}$$

comprise a nonlinear elliptic eigenvalue problem and the unknown eigenvalue μ is determined by the third equality,

$$\frac{E}{M^2} = \frac{\|\nabla v\|_2^2}{\left(\int_\Omega - \frac{\partial v}{\partial v}\right)^2}. \tag{47}$$

The elliptic theory ensures rather deailed features of the set of solutions to (46). Here we note the following facts [31].

1. There is $\bar\mu = \mu(\Omega) > 0$ such that the problem (46) does not admit a solution for $\mu > \bar\mu$.

2. Each $\mu \leq 0$ admits a unique solution.

3. Each $0 < \delta < \bar\mu$ admits a constant $C = C(\delta) > 0$ such that $\|v\|_\infty \leq C$ for any solution $v = v(x)$.

4. There is a family of solutions $\{(\mu, v)\}$ such that $\mu \downarrow 0$ and $\|v\|_\infty \to +\infty$.

We show the following theorem, consistent to Theorem 2.
Theorem 5 *If $\Omega = B(0, 1) \subset \mathbf{R}^2$, there is $\delta > 0$ such that any solution (v, μ) to (45) admits*

$$\frac{E}{M^2} \geq \delta. \tag{48}$$

Proof: If $\mu = 0$, it holds that $v = 0$. We have $\pm v > 0$ exclusively in Ω, provided that $\pm \mu > 0$, respectively. By the elliptic theory [32], therefore, any solution v to (46) is radially symmetric as in $v = v(r), r = |x|$. We have, furthermore, $\pm v_r < 0$ in $0 < r \leq 1$, if $\pm \mu > 0$, respectively.
Then it holds that $\psi = \psi(r)$, and hence

$$-\frac{1}{r}(r\psi_r)_r = \omega \quad \text{in } 0 < r \leq 1, \quad \psi|_{r=1} = 0 \tag{49}$$

by (42) and (43), which implies

$$-r\psi_r(r) = \int_0^r s\omega(s)ds > 0, \quad 0 < r \leq 1. \tag{50}$$

We thus obtain $\mu \neq 0$, in particular.
If $\mu < 0$ we have $\beta > 0$ by (44), and therefore, $\psi_r > 0$ in $0 < r \leq 1$ by $v_r > 0$ there. It is a contradiction, and hence $\mu > 0$. In this case, the solution $v = v(r)$ to (46) is explicit [31]. The numbers of the solution is 0, 1, and 2, according to $\mu > 2$, $\mu = 2$, and $0 < \mu < 2$, respectively, and if $0 < \mu \leq 2$ the solutions $v = v_\pm$ are given as

$$v_\pm(r) = \log \frac{8\gamma_\pm r}{(1 + \gamma_\pm r^2)^2}, \quad \gamma_\pm = \frac{4}{\mu}\left\{1 - \frac{\mu}{4} \pm \left(1 - \frac{\mu}{2}\right)^{1/2}\right\}. \tag{51}$$

In fact, we have $\gamma_+ = \gamma_-$ for $\mu = 2$.
This solution is parametrized by

$$\sigma = \int_\Omega \mu e^v \in (0, 8\pi). \tag{52}$$

Hence each $0 < \sigma < 8\pi$ admits a unique solution (v, μ) to (46), and $v = v_+$ and $v = v_-$ according as $\sigma \geq 4\pi$ and $\sigma \leq 4\pi$, respectively. It holds also that $\mu \downarrow 0$ if either

$\sigma \uparrow 8\pi$ or $\sigma \downarrow 0$. Thus we have only to confirm that E/M^2 is bounded, both as $\sigma \uparrow 8\pi$ and $\sigma \downarrow 0$.

As $\sigma \uparrow 8\pi$, we have

$$v = v_+(x) \quad \rightarrow \quad 4 \log \frac{1}{|x|} \quad \text{locally uniformly on } \overline{\Omega} \backslash \{0\} \tag{53}$$

and hence

$$\|\nabla v\|_2^2 \rightarrow +\infty, \quad \int_{\partial\Omega} -\frac{\partial v}{\partial \nu} \rightarrow 8\pi, \tag{54}$$

which implies

$$\lim_{\sigma \uparrow 8\pi} \frac{E}{M^2} = +\infty. \tag{55}$$

As $\sigma \downarrow 0$, on the other hand, we have

$$v = v_-(x) \quad \rightarrow \quad 0 \quad \text{uniformly in } \overline{\Omega}. \tag{56}$$

Since $\mu \downarrow 0$, furthermore, there arises that

$$\gamma = \gamma_- = \frac{4}{\mu}\left\{1 - \frac{\mu}{4} - \left(1 - \frac{\mu}{2}\right)^{1/2}\right\} = \mu(1 + o(1)). \tag{57}$$

It holds also that

$$v(r) = \log \frac{8\gamma}{\mu} - 2 \log\left(1 + \mu r^2\right) \tag{58}$$

and hence

$$v_r(r) = -\frac{4\mu r}{(1 + \mu r^2)^2} = -4\mu r(1 + o(1)) \quad \text{uniformly on } \overline{\Omega}. \tag{59}$$

Then, (59) implies

$$\|\nabla v\|_2^2 = 2\pi \int_0^1 v_r^2 r \ dr = 2\pi \cdot 16\mu^2 \cdot \int_0^1 r^3 \ dr \cdot (1 + o(1))$$
$$= 8\pi\mu^2(1 + o(1)) \tag{60}$$

as well as

$$\left(\int_{\partial\Omega} -\frac{\partial v}{\partial \nu}\right)^2 = 16\mu^2 \cdot 2\pi(1 + o(1)). \tag{61}$$

It thus follows that

$$\lim_{\sigma \downarrow 0} \frac{E}{M^2} = \frac{1}{4} \tag{62}$$

and hence the conclusion. $\qquad\qquad\qquad\qquad\qquad\qquad\qquad\qquad\qquad\qquad\square$

3. Proof of Theorem 1

The first observation is the following lemma.
Lemma 1 *Under the assumption of* (22), *it holds that*

$$\beta = \beta(t) < 0, \quad \omega_r(r,t) < 0, \quad 0 < r \leq 1, \quad 0 \leq t < T. \tag{63}$$

Proof: We have (7) and hence

$$\psi_r(r,t) < 0, \quad 0 < r \leq 1, \quad 0 \leq t < T \tag{64}$$

by (49), which implies, in particular,

$$\beta = -\frac{(\nabla\omega, \nabla\psi)}{\int_\Omega \omega|\nabla\psi|^2} < 0 \tag{65}$$

at $t = 0$ by (22).
Since $\omega = \omega(r,t)$ and $\psi = \psi(r,t)$, we obtain $\nabla^\perp\psi = 0$, and hence

$$\omega_t = \omega_{rr} + \frac{1}{r}\omega_r + \beta\psi_r\omega_r - \beta\omega^2 \tag{66}$$

by (1). Then $z = \omega_r$ satisfies

$$z_t = z_{rr} - \frac{1}{r^2}z + \frac{1}{r}z_r + \beta\psi_{rr}z + \beta\psi_r z_r - 2\beta\omega z, \quad 0 < r \leq 1, \quad 0 \leq t < T \tag{67}$$

$$z|_{r=0} = 0, \quad z|_{t=0} = \omega_{0r}(r) < 0, \quad 0 < r \leq 1$$

and

$$z = -\beta\omega\psi_r, \quad r = 1, \quad 0 \leq t < T. \tag{68}$$

Putting

$$m(t) = \min_{\partial\Omega} z(\cdot,t) = \omega_r(\cdot,t)|_{r=1}, \tag{69}$$

we obtain $m(0) < 0$ from the assumption. If there is $0 < t_0 <$ such that

$$m(t) < 0, \quad 0 \leq t < t_0 < T, \quad m(t_0) = 0, \tag{70}$$

we obtain $z(r,t) > 0$ for $0 \leq t < t_0, 0 < r \leq 1$, and $t = t_0, 0 < r < 1$ by the strong maximum principle. By (64), we have (65) for $0 \leq t \leq t_0$, that is,

$$\beta = -\frac{\int_0^1 \psi_r z r\, dr}{\int_0^1 \omega\psi_r^2 r\, dr} < 0, \quad 0 \leq t \leq t_0, \tag{71}$$

and hence

$$z = -\beta\omega\psi_r < 0 \quad r = 1, \quad t = t_0, \tag{72}$$

a contradiction. It holds that $z = \omega_r < 0$ for $0 \leq t < T, r = 1$, and hence

$$\beta = -\frac{\int_0^1 \psi_r \omega_r r\, dr}{\int_0^1 \omega\psi_r^2 r\, dr} < 0, \quad 0 \leq t < T. \quad \square \tag{73}$$

The proof of Theorem 3 relies on the fact

$$\beta \ge -C, \quad \int_\Omega \omega(\log\omega - 1) \le C \Rightarrow T = +\infty, \quad \|\omega(\cdot,t)\|_\infty \le C. \tag{74}$$

This property is known for the Smoluchoski-Poisson equation (34), but the proof is valid even to (31) with vortex term. Having (21), therefore, we have to provide the inequality $\beta \ge -C$.

The inequality $\beta < 0$, on the other hand, is sufficient for the following arguments.

Lemma 2 *If $\beta \le 0$, $0 \le t < T$, it holds that*

$$\omega \ge \underline{\omega} \equiv \min_{\overline{\Omega}} \omega_0 > 0 \quad on \ \overline{\Omega} \times \ [0,T). \tag{75}$$

Proof: Since (17) we obtain

$$\omega_t + \nabla^\perp\psi \cdot \nabla\omega = \Delta\omega + \beta\nabla\psi \cdot \nabla\omega + \beta\Delta\psi$$

$$= \Delta\omega + \beta\nabla\psi \cdot \nabla\omega - \beta\omega^2 \tag{76}$$

$$\ge \Delta\omega + \beta\nabla\psi \cdot \nabla\omega \quad in \ \Omega \times \ (0,T)$$

with

$$-\frac{\partial\omega}{\partial\nu} = \beta\omega\frac{\partial\psi}{\partial\nu} > 0 \quad on \ \partial\Omega \times \ [0,T) \tag{77}$$

by (8). Then the result follows from the comparison theorem. □

Lemma 3 *Under the assumption of the previous lemma, there is $C_0 = C_0(\Omega) > 0$ such that*

$$C_0\|\omega_0\|_2^3 \le E\underline{\omega} \Rightarrow \|\omega(\cdot,t)\|_2 \le \|\omega_0\|_2, \quad -\beta(t) \le \alpha \equiv \frac{\|\omega_0\|_2^2}{E\underline{\omega}}, \quad 0 \le t < T. \tag{78}$$

Proof: Using (11) and (17), we obtain

$$\int_\Omega [\nabla \cdot (\omega\nabla^\perp\psi)]\omega = \int_\Omega \omega\nabla\omega \cdot \nabla^\perp\psi = \frac{1}{2}\int_\Omega \nabla\omega^2 \cdot \nabla^\perp\psi$$

$$= -\frac{1}{2}\int_\Omega \omega^2\nabla \cdot \nabla^\perp\psi = 0. \tag{79}$$

Hence (1) with (2) implies

$$\frac{1}{2}\frac{d}{dt}\|\omega\|_2^2 + \|\nabla\omega\|_2^2 = -\beta\int_\Omega \omega\nabla\psi \cdot \nabla\omega = -\frac{\beta}{2}(\nabla\psi,\nabla\omega^2)$$

$$= -\frac{\beta}{2}\int_{\partial\Omega} \omega^2\frac{\partial\psi}{\partial\nu} + \frac{\beta}{2}(\Delta\psi,\omega^2) \le -\frac{\beta}{2}\|\omega\|_3^3 \tag{80}$$

by $\beta \le 0$ and (88). Since

$$\int_\Omega \nabla\omega \cdot \nabla\psi = \int_{\partial\Omega} \omega\frac{\partial\psi}{\partial\nu} + \int_\Omega \omega(-\Delta\psi) \le \int_\Omega \omega^2 \tag{81}$$

follows from (8), furthermore, it holds that

$$-\beta = \frac{\int_\Omega \nabla\omega \cdot \nabla\psi}{\int_\Omega \omega |\nabla\psi|^2} \le \underline{\omega}^{-1} \cdot \frac{\|\omega\|_2^2}{\|\nabla\psi\|_2^2} = \frac{1}{E\underline{\omega}} \|\omega\|_2^2. \tag{82}$$

Then ineqality (80) induces

$$\frac{1}{2}\frac{d}{dt}\|\omega\|_2^2 + \|\nabla\omega\|_2^2 \le \frac{1}{2E\underline{\omega}} \|\omega\|_2^2 \cdot \|\omega\|_3^3. \tag{83}$$

Here we use the Gagliardo-Nirenberg inequality (see (4.16) of [19]) in the form of

$$\|\omega\|_3^3 \le C\|\omega\|_{H^1} \cdot \|\omega\|_2^2 = C\|\omega\|_2^2 (\|\nabla\omega\|_2 + \|\omega\|_2), \tag{84}$$

to obtain

$$\frac{1}{2}\frac{d}{dt}\|\omega\|_2^2 + \|\nabla\omega\|_2^2 \le \frac{C}{E\underline{\omega}} \|\omega\|_2^4 (\|\nabla\omega\|_2 + \|\omega\|_2)$$

$$\le \frac{1}{2}\|\nabla\omega\|_2^2 + \frac{C^2}{8(E\underline{\omega})^2} \|\omega\|_2^8 + \frac{C}{2E\underline{\omega}} \|\omega\|_2^5 \tag{85}$$

and hence

$$\frac{d}{dt}\|\omega\|_2^2 + \|\nabla\omega\|_2^2 \le \frac{C}{E\underline{\omega}} \|\omega\|_2^5 \left(\frac{C}{E\underline{\omega}} \|\omega\|_2^3 + 1\right). \tag{86}$$

Then, Poincaré-Wirtinger's inequality ensures

$$\frac{d}{dt}\|\omega\|_2^2 + \mu\|\omega\|_2^2 \le \frac{C}{E\underline{\omega}} \left(\frac{C}{E\underline{\omega}} \|\omega\|_2^6 + \|\omega\|_2^3\right) \|\omega\|_2^2, \tag{87}$$

where $\mu = \mu(\Omega) > 0$ is a constant.
Writing

$$y(t) = \frac{C}{E\underline{\omega}} \|\omega\|_2^3, \tag{88}$$

we obtain

$$\frac{d}{dt}\|\omega\|_2^2 + \mu\|\omega\|_2^2 \le (y^2 + y)\|\omega\|_2^2, \tag{89}$$

and therefore, if

$$y^2 + y < \mu/2 \tag{90}$$

holds at $t = 0$, it keeps to hold that

$$\frac{d}{dt}\|\omega\|_2^2 \le 0 \tag{91}$$

and (90) for $0 \le t < T$. Then, we obtain

$$\|\omega(\cdot,t)\|_2 \le \|\omega_0\|_2, \quad 0 \le t < T, \tag{92}$$

and hence

$$-\beta(t) \le \frac{\|\omega_0\|_2^2}{E\underline{\omega}} = \alpha, \quad 0 \le t < T \tag{93}$$

by (82).

The condition $y(0) < \frac{\mu}{2}$ means

$$C_0 \|\omega_0\|_2 \le E\underline{\omega} \tag{94}$$

for $C_0 > 0$ sufficiently large, and hence we obtain the conclusion. □

Proof of Theorem 1: By the parabolic regularity, it suffices to show that

$$\|\omega(\cdot,t)\|_\infty \le C, \quad 0 \le t < T \tag{95}$$

under the assumption. We have readily shown

$$\|\omega(\cdot,t)\|_2 \le C, \; 0 \le -\beta(t) \le C, \quad 0 \le t < T \tag{96}$$

by Lemma 3. Then, the conclusion (95) is obtained similarly to (34). See [26] for more details.

In fact, we have

$$\int_\Omega [\nabla \cdot (\omega \nabla^\perp \psi)]\omega^p = -\int_\Omega \omega \nabla^\perp \psi \cdot \nabla \omega^p = -p \int_\Omega \omega^p \nabla^\perp \psi \cdot \nabla \omega$$

$$= -\frac{p}{p+1}\int_\Omega \nabla^\perp \psi \cdot \nabla \omega^{p+1} = \frac{p}{p+1}\int_\Omega \omega^{p+1} \nabla \cdot (\nabla^\perp \psi) = 0 \tag{97}$$

for $p > 0$ by (11) and (34). Then it follows that

$$\frac{1}{p+1}\frac{d}{dt}\int_\Omega \omega^{p+1} + \frac{4p}{(p+1)^2}\|\nabla \omega^{\frac{p+1}{2}}\|_2^2 = -\beta \int_\Omega \omega \nabla \psi \cdot \nabla \omega^p$$

$$= -\beta \cdot \frac{p}{p+1}\int_\Omega \nabla \psi \cdot \nabla \omega^{p+1} \le -\beta \frac{p}{p+1}\int_\Omega \omega^{p+1}(-\Delta \psi)$$

$$= -\beta \frac{p}{p+1}\int_\Omega \omega^{p+1} \le C \int_\Omega \omega^{p+2} \tag{98}$$

by $\beta < 0$ and (8). Then, Moser's iteration scheme ensures (95) as in [33].

4. Proof of Theorem 2

We begin with the following lemma.

Lemma 4 *Under the assumption of (22), it holds that*

$$-\beta(t) \ge \delta, \; 0 \le t < T, \; M = \|\omega_0\|_1 > \frac{8\pi}{\delta} \Rightarrow T < +\infty \tag{99}$$

in (31), where $\delta > 0$ is a constant.

Proof: We have $\omega = \omega(r,t)$ and $\psi = \psi(r,t)$ for $r = |x|$ under the assumption, which implies $\nabla^{\perp}\psi = 0$. Then we obtain

$$\nabla \cdot \omega \nabla^{\perp}\psi = \nabla\omega \cdot \nabla^{\perp}\psi = 0 \tag{100}$$

by (17). It holds also that

$$\nabla \cdot (\omega\nabla\psi) = \nabla \cdot \left(\omega\psi_r \frac{x}{r}\right) = \left(\nabla \cdot \frac{x}{r}\right)\omega\psi_r + \frac{x}{r} \cdot \nabla(\omega\psi_r)$$
$$= \frac{1}{r}\omega\psi_r + (\omega\psi_r)_r = \frac{1}{r}(r\omega\psi_r)_r, \tag{101}$$

and therefore, there arises that

$$\omega_t = \frac{1}{r}(r\omega_r + \beta r\omega\psi_r)_r, \qquad \omega_r + \beta\omega\psi_r|_{r=1} = 0. \tag{102}$$

from (31).
Then (102) implies

$$\frac{d}{dt}\int_0^1 \omega r^3 \, dr = \int_0^1 \omega_t r^3 \, dr = \int_0^1 (r\omega_r + \beta r\omega\psi_r)_r r^2 \, dr$$
$$= -\int_0^1 2r^2(\omega_r + \beta\omega\psi_r) \, dr \tag{103}$$
$$= -2r^2\omega\big|_{r=0}^{r=1} + \int_0^1 4r\omega - 2\beta\omega\psi_r r^2 \, dr.$$

Here we use (50) derived from the Poisson part of (31), that is,

$$-r\psi_r(r,t) = A(r,t) \equiv \int_0^r s\omega(s,t)ds. \tag{104}$$

Putting

$$\lambda = \int_0^1 \omega r \, dr = \frac{M}{2\pi}, \tag{105}$$

we obtain

$$\frac{d}{dt}\int_0^1 \omega r^3 \, dr = -2\omega|_{r=1} + 4\lambda + 2\beta\int_0^1 AA_r \, dr$$
$$= -2\omega|_{r=1} + 4\lambda + \beta A^2\big|_{r=0}^{r=1} \tag{106}$$
$$= -2\omega|_{r=1} + 4\lambda + \beta\lambda^2$$
$$< 4\lambda\left(\beta + \frac{M}{8\pi}\right) \le 4\lambda\left(-\delta + \frac{M}{8\pi}\right).$$

Since $-\delta + \frac{M}{8\pi} < 0$, therefore, $T = +\infty$ is impossible, and we obtain $T < +\infty$. □
Lemma 5 *Under the assumption (22), there is $\delta > 0$ such that*

$$\frac{E}{M^2} < \delta, \quad \beta(t) \le 0, \quad 0 \le t < T \Rightarrow \beta(t) \le -\frac{1}{CE^{1/2}} \quad 0 \le t < T. \tag{107}$$

Proof: First, Lemma 1 implies

$$\omega \geq \omega_* \equiv \omega|_{r=1}. \tag{108}$$

Second, we have

$$\int_\Omega \nabla \psi \cdot \nabla \omega = \int_{\partial\Omega} \frac{\partial \psi}{\partial \nu} \omega + \int_\Omega (-\Delta \psi) \omega = \omega_* \int_{\partial\Omega} \frac{\partial \psi}{\partial \nu} + \|\omega\|_2^2$$

$$= \omega_* \int_\Omega \Delta \psi + \|\omega\|_2^2 = \|\omega\|_2^2 - \omega_* M, \tag{109}$$

and hence

$$-\beta = \frac{\int_\Omega \nabla \psi \cdot \nabla \omega}{\int_\Omega \omega |\nabla \psi|^2} = \frac{\|\omega\|_2^2 - \omega_* M}{\int_\Omega \omega |\nabla \psi|^2}. \tag{110}$$

Here, we use the Gagliardo-Nirenberg inequality in the form of

$$\|w\|_4^2 \leq C \|w\|_2 \|w\|_{H^1}, \tag{111}$$

which implies

$$\int_\Omega \omega |\nabla \psi|^2 \leq \|\omega\|_2 \|\nabla \psi\|_4^2 \leq C \|\omega\|_2 \|\nabla \psi\|_2 \|\nabla \psi\|_{H^1}$$

$$\leq C E^{1/2} \|\omega\|_2^2 \tag{112}$$

by the elliptic estimate of the Poisson equation in (2),

$$\|\psi\|_{H^2} \leq C \|\omega\|_2. \tag{113}$$

We have, on the other hand,

$$\omega_* M \leq \frac{M}{E} \int_\Omega \omega |\nabla \psi|^2 \tag{114}$$

by (110), and therefore,

$$-\beta \geq \frac{1}{C E^{1/2}} - \frac{E}{M} \geq \frac{1}{2 C E^{1/2}}, \tag{115}$$

provided that

$$\frac{E}{M^2} < \left(\frac{1}{2C}\right)^2. \tag{116}$$

Then the conclusion follows. □

Proof of Theorem 2: By Lemma 5, there is $\delta_0 >$ such that

$$\frac{E}{M^2} < \delta \Rightarrow -\beta \geq \frac{1}{C E^{1/2}} \equiv \delta_1, \tag{117}$$

and then, Lemma 4 ensures

$$M > \frac{8\pi}{\delta_1} \Rightarrow T < +\infty. \tag{118}$$

The assumption in (118) means

$$\frac{E}{M^2} < \left(\frac{1}{8\pi c}\right)^2, \tag{119}$$

and hence we obtain the conclusion. $\qquad\qquad\qquad\qquad\qquad\qquad\qquad\square$

Appendix Proof of Theorem 4

This theorem is valid to the general case of Ω and ω_0 without (22). We assume $\delta = 1$ without loss of generation, so that

$$\beta \leq -1. \tag{120}$$

We follow the argument [27] concerning (34) with the Poisson part replaced by (42) or (43). Thus we have to take case of the vortex term $\nabla \cdot \omega \nabla^\perp \psi$, time varying $\beta = \beta(t)$, and the Dirichlet boundary condition in (31).

We recall the cut-off function used in [34] (see also Chapter 5 of [19]). Hence each $x_0 \in \overline{\Omega}$ and $0 < R \leq 1$ admit $\varphi = \varphi_{x_0, R} \in C^2(\overline{\Omega})$ with

$$\left.\frac{\partial\varphi}{\partial\nu}\right|_{\partial\Omega} = 0, \quad 0 \leq \varphi \leq 1, \quad \varphi = 1 \text{ in } \Omega \cap B(x_0, R/2), \quad \varphi = 0 \text{ in } \Omega \setminus B(x_0, R),$$
$$\tag{121}$$

and

$$|\nabla\varphi| \leq CR^{-1}\varphi^{1/2}, \quad |\nabla^2\varphi| \leq CR^{-2}\varphi^{1/2}. \tag{122}$$

In more details, we take a cut-off function, denoted by ψ, satisfying (121), using a local conformal mapping, and then put $\varphi = \psi^4$.

Let

$$\varphi \in C^2(\overline{\Omega}), \quad \left.\frac{\partial\varphi}{\partial\nu}\right|_{\partial\Omega} = 0. \tag{123}$$

be given. First, we have

$$\frac{d}{dt}\int_\Omega \omega\varphi = \int_\Omega \omega\nabla^\perp\psi \cdot \nabla\varphi - (\nabla\omega + \beta\omega\nabla\psi) \cdot \nabla\varphi \; dx$$
$$\tag{124}$$
$$= \int_\Omega \omega\nabla^\perp\psi \cdot \nabla\varphi + \omega\Delta\varphi - \beta\omega\nabla\psi \cdot \nabla\varphi \; dx$$

by (11). It holds that

$$\int_{\Omega} \omega \nabla \psi \cdot \nabla \varphi = \iint_{\Omega \times \Omega} \omega(x,t)[\nabla_x G(x,x') \cdot \nabla \varphi(x)]\omega(x',t) \; dxdx'$$

$$= \iint_{\Omega\Omega} \omega(x,t)\varphi_{x_0,2R}(x')[\nabla_x G(x,x') \cdot \nabla \varphi(x)]\omega(x',t) \; dxdx'$$

$$+ \iint_{\Omega \times \Omega} \omega(x,t)\left(1 - \varphi_{x_0,2R}(x')\right)[\nabla_x G(x,x') \cdot \nabla \varphi(x)]\omega(x',t) \; dxdx'$$

$$= I + II.$$

$$(125)$$

Let, furthermore, $x_0 \in \Omega$ and $0 < R \ll 1$ in the above equality. Then,

$$\varphi = |x - x_0|^2 \varphi_{x_0,.R} \qquad (126)$$

satisfies the requirement (123).
It holds that

$$\nabla \varphi = 2(x - x_0)\varphi_{x_0,R} + |x - x_0|^2 \nabla \varphi_{x_0,R} \qquad (127)$$

and hence

$$|\nabla \varphi| \le C|x - x_0|\left(\varphi_{x_0,R} + |x - x_0|R^{-1}\varphi_{x_0,R}^{1/2}\right) \le C|x - x_0|\varphi_{x_0,R}^{1/2}. \qquad (128)$$

We obtain, furthermore,

$$|x' - x_0| \ge 2R, \quad |x - x_0| \le R \Rightarrow |x - x'| \ge R, \qquad (129)$$

and hence

$$|\nabla_x G(x,x')| \le CR^{-1} \qquad (130)$$

in this case. Then it follows that

$$|II| \le CR^{-1}M \int_{\Omega} |x - x_0|\varphi_{x_0,R}^{1/2}\omega(x,t) \; dx \le CR^{-1}M^{3/2}A^{1/2}, \qquad (131)$$

where

$$A = \int_{\Omega} |x - x_0|^2 \varphi_{x_0,R}\omega. \qquad (132)$$

We have, on the other hand,

$$I = \iint_{\Omega \times \Omega} \omega(x,t)\varphi_{x_0,2R}(x')[\nabla_x G(x,x') \cdot \nabla \varphi(x)]\omega(x',t) \; dxdx'$$

$$= \frac{1}{2} \iint_{\Omega\Omega} [\varphi_{x_0,2R}(x')\nabla \varphi(x) \cdot \nabla_x G(x,x') + \varphi_{x_0,2R}(x)\nabla \varphi(x') \cdot \nabla_{x'} G(x,x')]\omega \otimes \omega,$$

$$(133)$$

where $G = G(x,x')$ is the Green's function to

$$-\Delta \psi = \omega, \quad \omega|_{\partial\Omega} = 0 \qquad (134)$$

and

$$\omega \otimes \omega = \omega(x,t)\omega(x',t) \; dxdx'. \tag{135}$$

Here we use the local property of the Green's function

$$G(x,x') = \Gamma(x-x') + K(x,x'), \quad K \in C^2\left(\overline{\Omega} \times \Omega\right) \cap C^2\left(\Omega \times \overline{\Omega}\right), \tag{136}$$

where

$$\Gamma(x) = \frac{1}{2\pi} \log \frac{1}{|x|} \tag{137}$$

stands for the fundamental solution to $-\Delta$.
Let

$$\rho^2_{x_0,R}(x,x()) = \varphi_{x_0,2R}(x')\nabla\varphi(x) \cdot \nabla_x K(x,x') + \varphi_{x_0,2R}\nabla\varphi(x') \cdot \nabla_{x'} K(x.x'). \tag{138}$$

Since (128) implies

$$|\varphi_{x_0,2R}(x')\nabla\varphi(x)| \le C\varphi_{x_0,2R}(x')|x - x_0|\varphi^{1/2}_{x_0,R}(x)$$

$$\le C|x - x_0|\varphi^{1/2}_{x_0,R}(x), \tag{139}$$

it holds that

$$|\rho^1_{x_0,R}(x,x')| \le C\left(|x - x_0|\varphi^{1/2}_{x_0,R}(x) + |x' - x_0|\varphi^{1/2}_{x_0,R}(x')\right). \tag{140}$$

Then, we obtain

$$I = \frac{1}{2}\iint_{\Omega\times\Omega} \rho^0_{x_0,R}(x,x')\omega \otimes \omega + III \tag{141}$$

with

$$|III| \le CM^{3/2}A^{1/2} \le CR^{-1}M^{3/2}A^{1/2}, \tag{142}$$

where

$$\rho^0_{x_0,R}(x,x') = \nabla\Gamma(x-x') \cdot \left(\varphi_{x_0,2R}(x')\nabla\varphi(x) - \varphi_{x_0,2R}(x)\nabla\varphi(x')\right). \tag{143}$$

Here, we have

$$\nabla\Gamma(x) = -\frac{x}{2\pi|x|^2}, \tag{144}$$

and therefore,

$$\rho^0_{x_0,R}(x,x') = \rho^2_{x_0,R}(x,x') + \rho^3_{x_0,R}(x,x') \tag{145}$$

fo

$$\rho^2_{x_0,R}(x,x') = -\frac{1}{2\pi}\frac{x-x'}{|x-x'|^2}\varphi_{x_0,2R}(x') \cdot (\nabla\varphi(x) - \nabla\varphi(x')) \tag{146}$$

$$\rho^3_{x_0,R}(x,x') = -\frac{1}{2\pi}\frac{x-x'}{|x-x'|^2}(\varphi_{x_0,2R}(x') - \varphi_{x_0,2R}(x)) \cdot \nabla\varphi(x). \tag{147}$$

Since (128) implies

$$|\rho_{x_0,R}^3(x,x')| \leq CR^{-1}|\nabla\varphi(x)| \leq CR^{-1}|x-x_0|\varphi_{x_0,R}^{1/2}(x), \tag{148}$$

there arises that

$$I = \frac{1}{2}\iint_{\Omega\times\Omega}\rho_{x_0,R}^2(x,x')\ \omega\otimes\omega + IV, \tag{149}$$

with

$$|IV| \leq CR^{-1}M^{3/2}A^{1/2}, \tag{150}$$

similarly.
We have, furthermore,

$$\nabla\varphi(x) - \nabla\varphi(x') = 2(x-x')\varphi_{x_0,R}(x) + 2(x'-x_0)(\varphi_{x_0,R}(x) - \varphi_{x_0,R}(x'))$$
$$+|x'-x_0|^2\left(\nabla\varphi_{x_0,R}(x) - \nabla\varphi_{x_0,R}(x')\right) + \left(|x-x_0|^2 - |x'-x_0|^2\right)\nabla\varphi_{x_0,R}(x), \tag{151}$$

and hence

$$\rho_{x_0,R}^2(x,x') = -\frac{1}{\pi}\varphi_{x_0,2R}(x')\varphi_{x_0,R}(x) + \rho_{x_0,R}^4(x,x') + \rho_{x_0,R}^5(x,x') + \rho_{x_0,R}^6(x,x') \tag{152}$$

with

$$|\rho_{x_0,R}^4(x,x')| \leq C|x-x'|^{-1}\varphi_{x_0,2R}(x')|x'-x_0||\varphi_{x_0,R}(x) - \varphi_{x_0,R}(x')|$$
$$\leq CR^{-1}|x'-x_0|\varphi_{x_0,2R}(x'), \tag{153}$$

$$|\rho_{x_0,R}^5(x,x')| \leq C|x-x'|^{-1}\varphi_{x_0,2R}(x')|x'-x_0|^2|\nabla\varphi_{x_0,R}(x) - \varphi_{x_0,R}(x')|$$
$$\leq CR^{-2}|x'-x_0|^2\varphi_{x_0,2R}(x') \tag{154}$$
$$\leq CR^{-1}|x'-x_0|\varphi_{x_0,2R}(x'),$$

and

$$|\rho_{x_0,R}^6(x,x')| \leq C|x-x'|\varphi_{x_0,2R}(x')||x-x_0|^2 - |x'-x_0|^2|\cdot|\nabla\varphi_{x_0,R}(x)|$$
$$\leq CR^{-1}(|x-x_0| + |x'-x_0|)\varphi_{x_0,R}(x)\varphi_{x_0,2R}(x') \tag{155}$$
$$\leq C(R^{-1}|x-x_0|\varphi_{x_0,R}(x) + R^{-1}|x'-x_0|\varphi_{x_0,2R}(x'))$$

by

$$||x-x_0|^2 - |x'-x_0|^2| = |(x-x',x+x'-2x_0)| \leq |x-x'|(|x-x_0| + |x'-x_0|). \tag{156}$$

The residual terms are thus treated similarly, and it follows that

$$\left|I + \frac{1}{2\pi}\int_\Omega \omega\varphi_{x_0,R}\cdot\int_\Omega \omega\varphi_{x_0,2R}\right| \leq CR^{-1}M^{3/2}A^{1/2}, \tag{157}$$

which results in

$$\left|\int_\Omega \omega\nabla\psi \cdot \nabla\varphi + \frac{1}{2\pi}\int_\Omega \omega\varphi_{x_0,R} \cdot \int_\Omega \omega\varphi_{x_0,2R}\right| \leq CR^{-1}M^{3/2}A^{1/2}. \tag{158}$$

We can argue similarly to the vortex term in (124). This time, from

$$\nabla^\perp\Gamma(x) \cdot x = 0 \tag{159}$$

it follows that

$$\left|\int_\Omega \omega\nabla^\perp\psi \cdot \nabla\varphi\right| \leq CR^{-1}M^{3/2}A^{1/2}. \tag{160}$$

Concerning the principal term of (124), we use

$$\Delta\varphi = 4\varphi_{x_0,R} + 4(x - x_0) \cdot \nabla\varphi_{x_0,R} + |x - x_0|^2\Delta\varphi_{x_0,R}. \tag{161}$$

From

$$|(x - x_0) \cdot \nabla\varphi_{x_0,R}| \leq CR^{-1}|x - x_0|\varphi_{x_0,R}^{1/2} \tag{162}$$

and

$$||-x_0|^2\Delta\varphi_{x_0,R}| \leq CR^{-2}|x - x_0|^2\varphi_{x_0,R}^{1/2}$$
$$\leq CR^{-1}|x - x_0|\varphi_{x_0,R}^{1/2}, \tag{163}$$

it follows that

$$\left|\int_\Omega \omega\Delta\varphi - 4\int_\Omega \omega\varphi_{x_0,R}\right| \leq C\int_\Omega R^{-1}|x - x_0|\varphi_{x_0,R}\omega \tag{164}$$
$$\leq CR^{-1}M^{1/2}A^{1/2}.$$

Let $M_1 = M_{x_0,R}$ and $M_2 = M_{x_0,2R}$ for

$$M_{x_0,R} = \int_\Omega \omega\varphi_{x_0,R}. \tag{165}$$

Then, using (120), we end up with

$$\frac{dA}{dt} \leq 4M_1 - \frac{M_1^2}{2\pi} + CR^{-1}\left(M^{3/2} + M^{1/2}\right)A^{1/2} + C(M_2 - M_1). \tag{166}$$

Inequalilty (166) implies $T < +\infty$ if $A(0) \ll 1$, as is observed by [27] (see also Chapter 5 of [19]). Here we describe the proof for completeness.

The first observation is the monotoniity formula

$$\left|\frac{d}{dt}\int_\Omega \omega\varphi\right| \leq C(M + M^2)\|\nabla\varphi\|_{C^1}, \tag{167}$$

derived from (124) and the symmetry of the Green's function: $G(x, x') = G(x', x)$. The proof is the same as in (34) and is omitted.

Second, we put $I_1 = I_{x_0,R}$ and $I_2 = I_{x_0,2R}$ for

$$I_{x_0,R} = \int_\Omega |x - x_0|^2 \omega \varphi_{x_0,R}. \tag{168}$$

Then it holds that

$$M_2 - M_1 \leq \int_{R < |x-x_0| < 2R} \varphi_{x_0,2R} \omega$$

$$\leq 2R^{-1} \int_\Omega |x - x_0| \varphi_{x_0,2R} \omega \leq 2M^{1/2} R^{-1} I_2^{1/2} \tag{169}$$

and

$$A_2 = A_1 + \int_\Omega |x - x_0|^2 \left(\varphi_{x_0,2R} - \varphi_{x_0,R} \right) \omega$$

$$\leq A_1 + 4R^2 \int_\Omega \left(\varphi_{x_0,2R} - \varphi_{x_0,R} \right) \omega, \tag{170}$$

which implies

$$\frac{dA_1}{dt} \leq 4M_1 - \frac{M_1^2}{2\pi} + CR^{-1} \left(M^{3/2} + M^{1/2} \right) A_1^{1/2}$$

$$+ C \left(M^{3/2} + M^{1/2} \right) \left\{ \int_\Omega \left(\varphi_{x_0,2R} - \varphi_{x_0,R} \right) \omega \right\}^{1/2}. \tag{171}$$

Here, we use (167) to ensure

$$\left| \frac{d}{dt} \left(4M_1 - \frac{M^2}{2\pi} \right) \right| \leq C(M + M^2) R^{-2} \tag{172}$$

and

$$\left| \frac{d}{dt} \int_\Omega \left(\varphi_{x_0,2R} - \varphi_{x_0,R} \right) \omega \right| \leq C(M + M^2) R^{-2}. \tag{173}$$

Then, it follows that

$$4M_1 - \frac{M_1^2}{2\pi} \leq 4M_1(0) - \frac{M_1(0)^2}{2\pi} + CBa \left(R^{-1} t^{1/2} \right) \tag{174}$$

and

$$\int_\Omega \left(\varphi_{x_0,2R} - \varphi_{x_0,R} \right) \omega \leq \int_\Omega \left(\varphi_{x_0,2R} - \varphi_{x_0,R} \right) \omega_0 + CBa \left(R^{-1} t^{1/2} \right)$$

$$\leq 2R^{-2} A_2(0) + CBa \left(R^{-1} t^{1/2} \right) \tag{175}$$

for

$$B = M^{3/2} + M^{1/2}, \quad a(s) = s^2 + s. \tag{176}$$

Thus we obtain

$$\frac{dA_1}{dt} \le 4M_1(0) - \frac{M_1(0)^2}{2\pi} + CR^{-1}BA_1^{1/2} + CBA_2(0)^{1/2} + CBa\left(R^{-1}t^{1/2}\right)$$

$$= J(0) + CBa\left(R^{-1}t^{1/2}\right) + CBR^{-1}A_1^{1/2} \tag{177}$$

for

$$J = 4M_1 - \frac{M_1^2}{4\pi} + CBR^{-1}A_2^{1/2}. \tag{178}$$

Assume $M_1(0) > 8\pi$, and put

$$-4\delta = 4M_1(0) - \frac{M_1(0)^2}{2\pi} < 0. \tag{179}$$

Let, furthemore,

$$\frac{1}{R^2} \int_\Omega |x - x_0|^2 \varphi_{x_0,2R} \omega_0 \le \eta. \tag{180}$$

Now we define s_0 by

$$CBa(s_0) = \delta \tag{181}$$

in (177), and take $0 < \eta \ll 1$ such that

$$\eta \le \delta s_0^2. \tag{182}$$

Then, if R and T_0 satisfy $R^{-2}T_0 = \eta\delta^{-1}$, it holds that

$$A_1(0) \le R^2\eta < 2\delta T_0. \tag{183}$$

Making $0 < \eta \ll 1$, furthermore, we may assume

$$J(0) + CBR^{-1}A_1(0)^{1/2} \le -4\delta + CBR^{-1}A_2(0)^{1/2}$$

$$\le -4\delta + CB\eta^{1/2} \le -3\delta, \tag{184}$$

which results in

$$\frac{dA_1}{dt} \le J(0) + CBa\left(R^{-1}T_0^{1/2}\right) + BR^{-1}A_1(t)^{1/2}$$

$$= J(0) + \delta + CBR^{-1}A_1^{1/2}, \quad 0 \le t < T_0, \tag{185}$$

provided that $T \ge T_0$.
A continuation argument to (184)–(185) guarantees

$$\frac{dA_1}{dt} \le -2\delta, \quad 0 \le t < T_0, \tag{186}$$

and then we obtain

$$A_1(T_0) \leq A_1(0) - 2\delta T_0 < 0 \qquad (187)$$

by (183), a contradiction. $\qquad\qquad\qquad\qquad\qquad\qquad\qquad\qquad\qquad$ □

Acknowledgements

This work was supported by JSPS Grand-in-Aid for Scientific Research 19H01799.

Author details

Ken Sawada[1] and Takashi Suzuki[2]*

1 Meteorological College, Asahi-cho, Kashiwashi, Japan

2 Center for Mathematical Modeling and Data Science, Osaka University, Toyonakashi, Japan

*Address all correspondence to: suzuki@sigmath.es.osaka-u.ac.jp

IntechOpen

References

[1] P.H. Chavanis, *Generalized thermodynamics and Fokker-Planck equations: Applications to stellar dynamics and two-dimensional turbulence*, Phys. Rev. E68, (2003) 036108.

[2] P.-H. Chavanis, *Two-dimensional Brownian vortices*, Physica A 387 (2008) 6917-6942.

[3] P.H. Chavanis, J. Sommeria, R. Robert, *Statistical mechanics of two-dimensional vortices and collisionless stellar systems*, Astrophys. J. 471 (1996) 385-399.

[4] K. Sawada, T, Suzuki, *Relaxation theory for point vortices*, In; Vortex Structures in Fluid Dynamic Problems (H. Perez-de-Tejada ed.), INTECH 2017, Chapter 11, 205-224.

[5] D. Lynden-Bell, *Statistical mechanics of violent relaxation in stellar systems*, Monthly Notices of Royal Astronmical Society 136 (1967) 101-121.

[6] E. Caglioti, P.L. Lions, C. Marchioro, M. Pulvirenti, *A special class of stationary flows for two-dimensional Euler equations: a statistical mechanics description* Comm. Math. Phys. 143 (1992) 501.

[7] G.L. Eyink, H. Spohn, *Negative-temperature states and large scale, long-visited vortices in two-dimensional turbulence*, Statistical Physics 70 (1993) 833.

[8] G. Joyce, D. Montgomery, *Negative temperature states for two-dimensional guiding-centre plasma*, J. Plasma Phys. 10 (1973) 107.

[9] M.K.H. Kiessling, *Statistical mechanics of classical particles with logarithmic interaction*, Comm. Pure Appl. Math. 46 (1993) 27-56.

[10] K. Nagasaki, T. Suzuki, *Asymptotic analysis for two-dimensional elliptic eigenvalue problems with exponentially dominated nonlinearities*, Asymptotic Analysis 3 (1990) 173-188.

[11] L. Onsager, *Statistical hydrodynamics*, Suppl. Nuovo Cimento 6 (1949) 279-287.

[12] Y.B. Pointin, T.S. Lundgren, *Statistical mechanics of two-dimensional vortices in a bounded container*, Phys. Fluids 19 (1976) 1459-1470.

[13] R. Robert, J. Sommeria, *Statistical equilibrium states for two-dimensional flows*, J. Fluid Mech. 229 (1991) 291-310.

[14] R. Robert, J. Sommeria, *Relaxation towards a statistical equilibrium state in two-dimensional perfect fluid dynamics*, Phys. Rev. Lett. 69 (1992) 2776-2779.

[15] R. Robert, *A maximum-entropy principle for two-dimensional perfect fluid dynamics* J. Stat. Phys. **65** (1991) 531-553.

[16] R. Robert, C. Rosier, *The modeling of small scales in two-dimensional turbulent flows: A statistical Mechanics Approach*, J. Stat. Phys. 86 (1997) 481-515.

[17] K. Sawada, T. Suzuki, *Derivation of the equilibrium mean field equations of point vortex system and vortex filament system*, Theor. Appl. Mech. Japan 56 (2008) 285-290.

[18] C. Sire, P.-H. Chavanis, *Thermodynamics and collapse of self-gravitating Brownian particles in D dimensions*, Phys. Rev. E 66 (2002) 046133.

[19] T. Suzuki, *Free Energy and Self-Interacting Particles*, Birkhäuser, Boston, 2005.

[20] T. Suzuki, *Mean Field Theories and Dual Variation - Mathematical Structures of Mesoscopic Model*, 2nd edition, Atlantis Press, Paris, 2015.

[21] T. Suzuki, *Chemotaxis, Reaction, Network - Mathematics for Self-Organization*, World Scientific, Singapore, 2018.

[22] T. Suzuki, *Liouville's Theory in Linear and Nonlinear Partial Differential Equations - Interaction of Analysis, Geometry, Physics*, Springer, Berlin, (to appear).

[23] T. Suzuki, *Applied Analysis - Mathematics for Science, Engineering, Technology*, 3rd edition, Imperial College Press, London, (to appear).

[24] P. Biler, *Local and global solvability of some parabolic systems modelling chemotaxis*, Adv. Math. Sci. Appl. 8 (1998) 715-743.

[25] H. Gajewski, K. Zacharias, *Global behaviour of a reaction-diffusion system modelling chemotaxis*, Math. Nachr. 195 (1998) 77-114.

[26] T. Nagai, T. Senba, and K. Yoshida, *Application of the Trudinger-Moser inequality to a parabolic system of chemotaxis*, Funkcial. Ekvac. 40 (1997) 411-433.

[27] T. Senba and T. Suzuki, *Parabolic system of chemotaxis; blowup in a finite and in the infinite time*, Meth. Appl. Anal. 8 (2001) 349-368.

[28] W. Jäger, S. Luckhaus, *On explosions of solutions to a system of partial differential equations modelling chemotaxis*, Trans. Amer. Math. Soc. 329 (1992) 819-824.

[29] T. Nagai, *Blow-up of radially symmetric solutions to a chemotaxis system*, Adv. Math. Sci. Appl. 5 (1995) 581-601.

[30] T. Suzuki, *Exclusion of boundary blowup for 2D chemotaxis system provided with Dirichlet condition for the Poisson part*, J. Math. Pure Appl. 100 (2013) 347-367.

[31] T. Suzuki, *Semilinear Elliptic Equations - Classical and Modern Theories*, De Gruyter, Berlin, 2020.

[32] B. Gidas, W.M. Ni, L. Nirenberg, *Symmetry and related properties via the maximum principle*, Comm. Math. Phys. 68 (1979) 209-243.

[33] N.D. Alikakos, L^p *bounds of solutions of reaction-diffusion equations*, Comm. Partial Differential Equations 4 (1979) 827-868.

[34] T. Senba and T. Suzuki, *Chemotactic collapse in a parabolic-elliptic system of mathematical biology*, Adv. Differential Equations 6 (2001) 21-50.

Section 3

Vortex Phenomena in Planetary Atmospheres and Dusty Plasma

.

Chapter 7

Vortex Dynamics in the Wake of Planetary Ionospheres

Hector Pérez-de-Tejada and Rickard Lundin

Abstract

Measurements conducted with spacecraft around Venus and Mars have shown the presence of vortex structures in their plasma wake. Such features extend across distances of the order of a planetary radius and travel along their wake with a few minutes rotation period. At Venus, they are oriented in the counterclockwise sense when viewed from the wake. Vortex structures have also been reported from measurements conducted by the solar wind-Mars ionospheric boundary. Their position in the Venus wake varies during the solar cycle and becomes located closer to Venus with narrower width values during minimum solar cycle conditions. As a whole there is a tendency for the thickness of the vortex structures to become smaller with the downstream distance from Venus in a configuration similar to that of a cork-screw flow in fluid dynamics and that gradually becomes smaller with increasing distance downstream from an obstacle. It is argued that such process derives from the transport of momentum from vortex structures to motion directed along the Venus wake and that it is driven by the thermal expansion of the solar wind. The implications of that momentum transport are examined to stress an enhancement in the kinetic energy of particles that move along the wake after reducing the rotational kinetic energy of particles streaming in a vortex flow. As a result, the kinetic energy of plasma articles along the Venus wake becomes enhanced by the momentum of the vortex flow, which decreases its size in that direction. Particle fluxes with such properties should be measured with increasing distance downstream from Venus. Similar conditions should also be expected in vortex flows subject to pressure forces that drive them behind an obstacle.

Keywords: vortex Venus plasma wake, solar wind-Venus interaction, plasma acceleration in the Venus wake

1. Introduction

Measurements conducted with the Pioneer Venus Orbiter (PVO) and the Venus Express spacecraft (VEX) around Venus have provided evidence on the existence of vortex structures in the Venus plasma wake [1–5]. Much of what has been learned led to estimate the ($\sim 1\ R_V$) scale size of those features across the wake, which are shown in the left panel of **Figure 1** with a view of the velocity vectors projected on the plane transverse to the solar wind direction. The flow pattern corresponds to a vortex gyration of the velocity vectors of the solar wind H+ ions, which is also accompanied by a similar distribution of the velocity vectors of the planetary O+ ions that have been dragged by the solar wind from the Venus upper ionosphere. Comparative indications on the presence of vortex structures in the Mars plasma environment have also

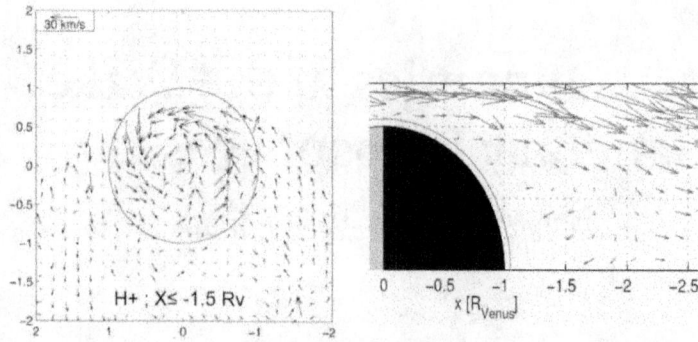

Figure 1.
(Left panel) Velocity vectors of H + ≈ 1–300 eV ions measured with the VEX spacecraft in the Venus near wake projected on the YZ plane transverse to the solar wind direction. Data are averaged in 1000 × 1000 km columns at X < −1.5 R_V (adapted from Figure 4 of [2]). (Right panel) Average direction of solar wind ion velocity vectors across the Venus near wake collected from many VEX orbits and projected in cylindrical coordinates [7].

been inferred from the observation of ionospheric-sheath boundary oscillations from the MAVEN plasma data [6]. In this case, measurements suggest Kelvin-Helmholtz oscillations with a periodicity of ~3 min but that have not yet completed a full vortex turn. At Venus, there is evidence of a reversal in the direction of the velocity vectors of plasma particles that lead them back to the planet in the central part of the Venus wake as is shown in the right panel of **Figure 1**.

Evidence of vortex structures in the Venus wake is also available from changes in the magnetic field direction in the VEX measurements. In previous reports [4, 8], it has been pointed out that at the time when a vortex structure is identified by a local increase of the plasma density in the Venus wake, there is an accompanying

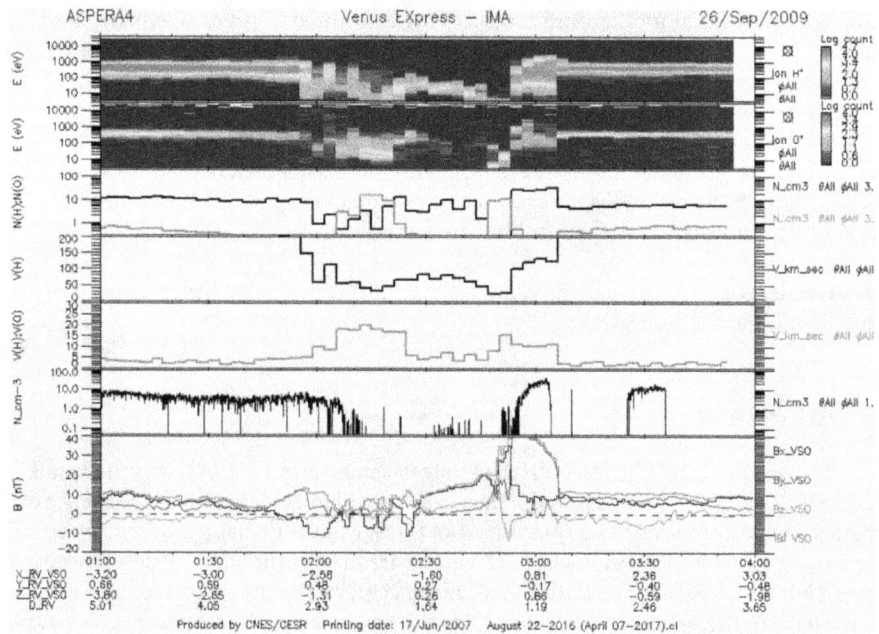

Figure 2.
Energy spectra of the H+ and O+ ions (upper panels) with measurements in the Sept 26–2009 VEX orbit where there are strong oscillations in the magnetic field components between 02:05 UT and 02:30 UT (bottom panel). In that time span there are enhanced density and speed values of the O+ ions (third and fifth panels).

decrease of the local magnetic field convected by the solar wind with distinct changes in the orientation of its components. A useful example with such properties is reproduced in **Figure 2** to show evidence of sudden changes in the orientation of the magnetic field components between 02:00 UT and 02:30 UT (bottom panel) when the plasma density and speed of planetary particles in the Venus wake show enhanced values (third and fifth panels). Despite the fact that the data has provided notable information on the general characteristics of vortex structures, there remains to address important aspects related to their origin and to the dynamics of their motion. These issues will be examined by considering the various fluid forces that are imposed by the solar wind on the planetary ions that stream in the wake. The description to be addressed here refers to the Venus observations where there is ample evidence on the geometry, distribution, and time variations of vortex structures. Despite the presence of small fossilized magnetic field areas in the Mars surface [9], much of the information for the Venus wake will apply to plasma vortices that stream along the Mars plasma wake.

2. Fluid dynamic forces in the Venus wake

A dominant feature in the motion of the solar wind particles that stream around the Venus ionosphere is that different from the pile-up of magnetic field fluxes convected by the solar wind over the dayside hemisphere, the plasma experiences local heating processes when it moves by the terminator of the Venus ionosphere. Indications of that plasma heating were first obtained from the Venera measurements through crossings of that spacecraft across the Venus wake with enhanced plasma temperature values along the flanks of the Venus ionosheath and that is reproduced in **Figure 3** [10, 11]. The heating derives from dissipation processes produced by the transport of solar wind momentum mostly over the Venus polar ionosphere where the local pile-up of the solar wind magnetic field fluxes is not strong. A suitable added information in the data of **Figure 3** is the presence of a plasma transition in the temperature and speed profiles (at ~02:00 Moscow time), which together with the vortex structure shown in **Figure 1** are unrelated to the dynamics expected from sling shot effects produced by the magnetic field fluxes draped around Venus [12]. While measurements have shown an antisolar directed motion of planetary ions in the Venus wake, a vortex flow configuration like that shown in **Figure 1** is more complex than that expected from a slingshot geometry.

As a result of the enhanced plasma temperature values, the solar wind expands by thermal pressure forces and then moves into the Venus wake from both polar regions. An implication of that displacement is that there are two separate flows of plasma particles reaching the central wake from two opposite directions along the Z-axis. They move from both polar regions where the planetary O+ ions are first subject to low values of the rotation of the Venus ionosphere, and then they are displaced to equatorial latitudes where the rotation speed of the Venus ionosphere is larger. Since both plasma flows are also streaming along the X-axis following the solar wind direction, there should be a Coriolis force that deflects them in opposite directions along the Y-axis. For both flows the deflection of the particles should not be in the same sense since in the northern hemisphere the particles will move in the –Z direction and in the southern hemisphere in the +Z-direction. In addition to such opposite deflection along the Y-axis, the streaming particles will also be influenced by the effects of the aberrated direction of the solar wind and a general Magnus force that drive all planetary ions in motion around the planet with a velocity component directed in the +Y sense ([13], see their Fig. 15; [14]). Since that effect is contrary to the direction of motion along the -Y sense imposed by the

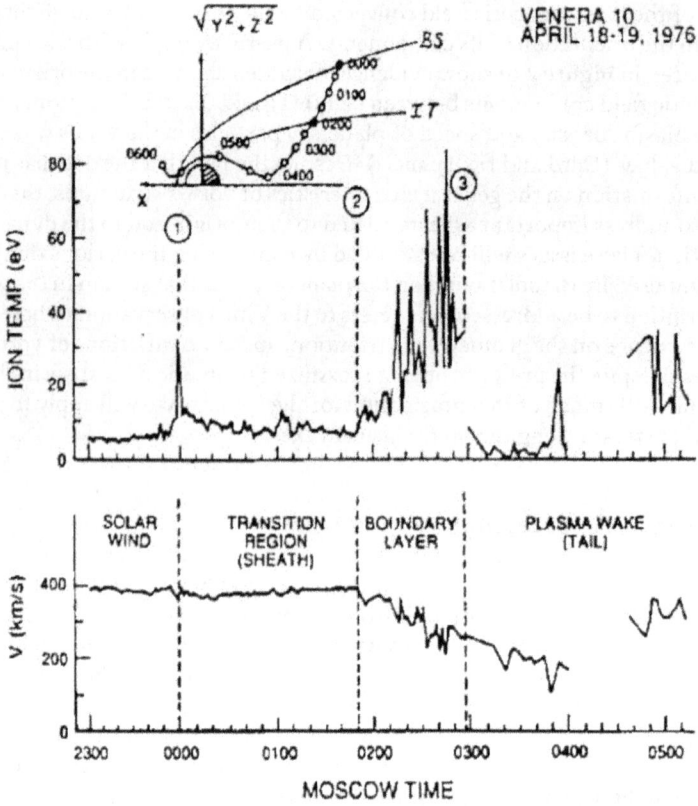

Figure 3.
Ion speed and temperature measured along the orbit of Venera 10 on Apr. 19, 1976. The Venera orbit in cylindrical coordinates is shown at the top. The temperature burst at position 1 was recorded during a flank crossing of the shock wave. The boundary layer is apparent by the increase in temperature and decrease in speed and is initiated by the intermediate transition at position labeled 2. The discontinuity in the boundary layer temperature profile corresponds to the boundary of the magneto-tail. (from [10]).

Coriolis force to the O+ ions that stream in the southern hemisphere, their resulting total velocity will be smaller than that of the O+ ions in the northern hemisphere where the velocity components implied by the Coriolis and the Magnus force are oriented in the same sense along the +Y direction.

An implication of that velocity difference between both hemispheres is that the momentum flux of the O+ ions along the Y-axis in the southern hemisphere will be smaller than the momentum flux of the O+ ions in the northern hemisphere. Consequently, a fraction of the momentum flux of the O+ ion fluxes that move north along the Z-axis in the southern hemisphere from the polar region (derived from the enhanced thermal pressure force at that region) will be transferred to that in the Y-sense to compensate for the smaller values of their momentum flux with respect to the larger Y-directed momentum flux of the O+ ions that stream in the northern hemisphere. Thus, in addition that the larger momentum flux of the O+ ions in the XY plane in the northern hemisphere over that of the O+ ions in the southern hemisphere, there will also be smaller values for the momentum flux of the O+ ions that move north along the Z-axis in the southern hemisphere. Under such conditions the momentum flux of the O+ ions that are directed south in the northern hemisphere will be dominant over that directed north in the southern hemisphere. As a result, the motion of the O+ ions in the northern hemisphere will force the vortex structure to be displaced toward the −Z direction. Such an effect

		Inbound						Outbound				
Date	UT	X	Y	Z	v	UT	X	Y	Z	v	δ	Δ
Aug 22/06	01.44	−2.90	−0.15	**−0.85**	25	01:54	−2.55	−0.15	**−0.40**	25	0.35	0.45
Aug 23/06	01:56	−2.75	−0.07	**−0.60**	15	02:05	−2.20	−0.07	**−0.20**	15	0.35	0.40
Aug 24/06	02:10	−2.45	−0.01	**−0.20**	30	02:20	−2.05	−0.02	**0.20**	30	0.40	0.40
Aug 28/06	02:22	−2.40	−0.28	**−0.35**	20	02:28	−1.90	0.20	**0.00**	20	0.50	0.35
Sep 19/09	01:54	−2.38	−0.04	**−0.80**	15	02:03	−2.11	−0.05	**−0.40**	15	0.30	0.40
Sep 21/09	02:02	−2.30	0.09	**−0.65**	15	02:12	−1.95	0.06	**−0.12**	15	0.35	0.53
Sep 25/09	02:14	−2.10	0.33	**−0.45**	20	02:27	−1.60	0.23	**0.25**	28	0.50	0.70
Sep 26/09	02:12	−2.30	0.42	**−0.70**	20	02:21	−1.95	**0.35**	**−0.20**	20	0.35	0.50

Table 1.
VEX coordinates along the X, Y, and Z-axis (in R_V) together with the speed v of the planetary O+ ions (in km/s), measured during the inbound (left columns) and the outbound (right columns) crossings of a vortex structure in selected VEX orbits across the Venus wake (the last two columns are the extent of the vortex structure measured in the X and in the Z directions in R_V).

agrees with the data of the VEX position of the vortex structures measured in the XZ plane during the 2006 and 2009 orbits listed in **Table 1** and that is also represented in their profiles in **Figure 4** showing how they become directed to lower −Z values with increasing distance downstream from Venus ([8], see their **Figure 3**). As a result, the outcome of the different momentum flux of the planetary O+ ions in the northern and in the southern hemispheres in the Venus wake is related to the

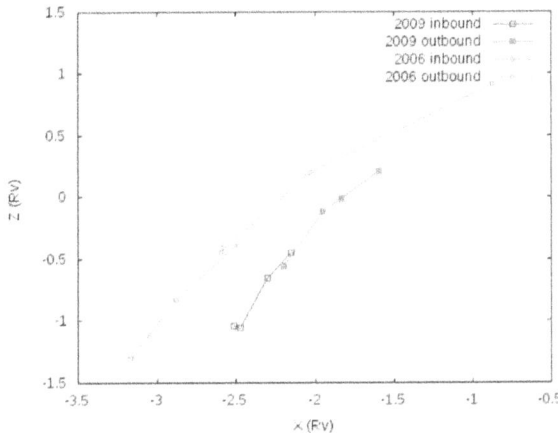

Figure 4.
*Position of the VEX spacecraft projected on the XZ plane during its entry (inbound) and exit (outbound) through a corkscrew plasma structure in several orbits. The two traces correspond to four orbits in 2006 and four orbits in 2009 listed in **Table 1** [8].*

effects produced by the Coriolis and the Magnus force and that lead to the south-bound displacement of the vortex structures shown in **Figure 4**.

3. Cross section of the vortex structures along the Venus wake

Data corresponding to the 2006–2009 orbits discussed in **Figure 4** have been further addressed to examine the shape of the vortex structures along the Venus wake with results that are presented in **Table 1**. We have separately collected information obtained for the inbound (left side) and for the outbound (right side) columns in each set of orbits. Values for the extent between both crossings along the X and Z-axis are indicated in the two last columns.

A notable aspect of the data in **Table 1** is that the values of the X-coordinate for the inbound and the outbound crossings of the four orbits during 2006 are larger than those for 2009. The implication is that the vortex structure is located closer to Venus during conditions approaching the solar cycle minimum by 2009. The same conclusion has also been inferred from the relative position of the 2006 and 2009 profiles in **Figure 4**. At the same time, the values of the X-coordinate during the inbound crossings in all eight orbits of **Table 1** (left side columns) are larger than those of the outbound crossings. Such difference derives from the tilted orientation of the trajectory of the VEX spacecraft on the XZ plane, which is directed toward Venus from the wake as it moves from the southern to the northern hemispheres (the inbound crossing of VEX is encountered at a larger distance downstream from Venus in the southern hemisphere as it then moves to a closer distance to Venus during its outbound crossing in the northern hemisphere).

A detail calculation has now been conducted to the data in the orbits of **Table 1** to estimate the width and the location of the vortex structures measured in both the 2006 and the 2009 orbits. The results are shown in **Figure 5** and indicate that there is a tendency for the vortex structures in the 2006 orbits to occur farther away from Venus than those in the 2009 orbits. At the same time there is an indication that the time width ΔT between the inbound and the outbound encounters marked by the vertical coordinate in **Figure 5** occurs at smaller values in the 2006 orbits, which trace the wake farther away from Venus. Such would be the case in a corkscrew flow configuration in fluid dynamics where its width becomes smaller with downstream distance from an obstacle as is represented in **Figure 6**.

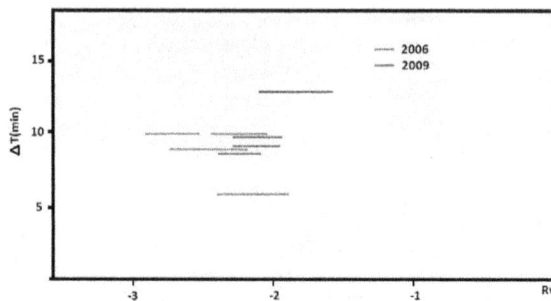

Figure 5.
*Segments measured between the inbound and the outbound crossings of vortex structures by VEX in the eight orbits included in **Table 1** (they are identified by changes in the particle flux intensity measured in the energy spectra of the O+ ions). Their position along the X-axis shows that the 2009 orbits (marked in blue) are located closer to Venus and that the width of the 2006 orbits (marked in red) is smaller since they have lower ΔT values and are encountered further downstream along the wake.*

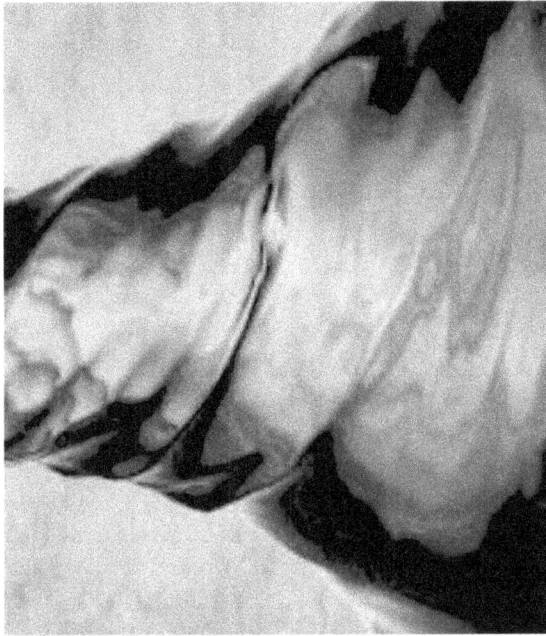

Figure 6.
View of a corkscrew vortex flow in fluid dynamics. Its geometry is equivalent to that of a vortex flow in the Venus wake with its width and position varying during the solar cycle. Near minimum solar cycle conditions, the vortex is located closer to Venus (right side) and there are indications that its width becomes smaller with increasing distance downstream from the planet. Such is the case for the 2006 orbits (marked in red) traced in **Figure 5** *and that were conducted before the solar cycle minimum at 2009–2010, thus implying that the vortex flow becomes thinner when it is detected further downstream along the wake.*

A view that can be proposed from such difference is that the width of the vortex structures is larger during the 2009 orbits (when they are located closer to Venus) rather than in 2006 (when they occurred further downstream). A general description of vortex structures measured in the Venus wake is that as shown in **Figure 4,** they are formed closer to Venus during minimum solar cycle conditions by 2009 and at the same time they are wide features. Different properties are encountered as the solar cycle progresses since they will now be formed further downstream from Venus and as shown in **Table 1** and in **Figure 5,** they will now extend across a smaller cross section within the wake. More extended calculations are still required to examine the geometric properties of the vortex structures listed in the 20 VEX orbits reported by Pérez-de-Tejada and Lundin [8]. In particular, it will be necessary to address whether the width of the vortex structures becomes narrower when they are measured further downstream from Venus. It was pointed out in that report that the width of the vortex structures becomes narrower when they are measured further downstream from Venus.

Thus, the thickness of the vortex structures gradually decreases with distance downstream from Venus and eventually fade away and diffuse with the solar wind plasma. Further studies of more extended data are required to examine the evolution of the vortex structures far downstream along the Venus tail. A view of the distribution of vortex structures along the Venus wake as they follow the motion of plasma particles can be inferred by comparing their cross section formed between the inbound and the outbound crossings on the XZ plane for the different orbits in **Table 1**. The result of that comparison is presented in **Figure 7** where the width "Δ" of the vortex structures along the Z-axis is compared with that of their extent "δ"

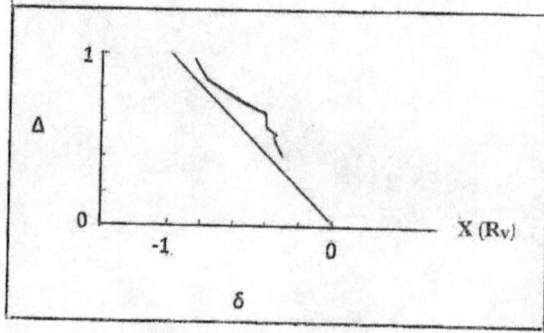

Figure 7.
Values of the extent "δ" of vortex structures along the wake (X-axis) and their width "Δ" along the Z-axis as derived from the data in **Table 1.** *Thin features (small "Δ" values) are obtained in the 2009 orbits while wider vortices correspond to the 2006 orbits.*

along the X-axis. It is notable that this latter quantity is correlated with their width along the Z-axis implying that thin vortices have a shorter extent along the wake and that wide vortices have a larger extent in that direction. A peculiar characteristic of the "δ" and the "Δ" values in the trace shown in **Figure 7** is that $\Delta > \delta$ (by comparison, a linear relation between them is shown by the straight line for the case in which they have the same value). Thus, as is indicated in the last two columns of **Table 1,** there is a tendency that along the VEX trajectory the vortex structures have a larger width along the Z-axis (that difference may be due to the tilted orientation of the VEX trajectory on the XZ plane or to enhanced Δ values produced by an asymmetric shape of the vortex structure in that plane).

4. Velocity values of plasma particles along the Venus wake

In **Table 1,** there is evidence that the speed values of the planetary O+ ions vary between 15 and 30 km/s by the inbound and the outbound crossings of vortex structures in the 2006 and 2009 orbits. Such values correspond to measurements conducted along the sun-Venus direction (X-axis) and thus are not related to changes produced by the vortex motion whose speed values vary across the wake. Vortex motion involves speed values of the order of ~200 km/s, which are derived from the transit time of particles around structures with a 1 R_V planetary radius during a ~ 3 min rotation period T [6]. With such high-speed values the plasma particles contain a large fraction of the momentum flux brought in by the solar wind and that has been employed to produce the vortex motion that they follow within the wake. As noted above, the vortex features are displaced in a consistent unified manner along the wake with much smaller speed values.

In addition to the different speed of the vortex structures and that of the particles that move within them, it should be noted that vortex motion marks a response different from that expected from motion produced by the convective electric field of the solar wind. Rather than following the directional motion of the solar wind with a gyrotropic trajectory at nearly the same speed as is predicted in that view the available momentum flux is employed to produce, instead, an alternate vortex flow configuration that is displaced coherently with more moderate speeds.

Despite the fact that there is no indication in the data of **Table 1** on the manner in which the speed values of the vortex structure change with distance along the Venus wake, it is possible to obtain that information from the varying values of the width "Δ" of the vortex structures that is implied from those obtained during 2006 and 2009. From the data in **Table 1** we can assume $\Delta = 0.765\ R_V$ as the average value for the width of the vortex structures during the 2006 orbits and $\Delta = 0.532\ R_V$ for that obtained in the 2009 orbits. Since such change implies a scale size decrease by a factor of 0.50% in the area "A" of the vortex structures, we can apply conservation of mass flux $\rho v^2 A = cst$ to suggest that their speed will increase to nearly double values assuming that the plasma density remains the same.

By applying this procedure, it is possible to argue that as the vortex structures become narrower with downstream distance from Venus, the kinetic energy of planetary particles that stream along the wake will be increased as a result of the decreased values of the cross section of the vortex structures. Thus, it should not be unexpected to measure higher-speed planetary ions moving in the far reaches of the Venus wake.

While most measurements of the vortex structures reported in **Table 1** are applicable to conditions that occur near the midnight plane of the Venus wake (small values of the Y-coordinate in the data of **Figure 2**), there are a few cases (orbits Sept 25–26 in 2009 and Aug 28 in 2006 in **Table 1**) where such structures are encountered at large distances away from that plane; that is, with larger Y-values. Those features also involve enhanced values of the O+ ion density values separated from the Venus ionosphere and that are included in the corresponding panels of figures like those shown in **Figure 2**. In particular, the location where the enhanced density

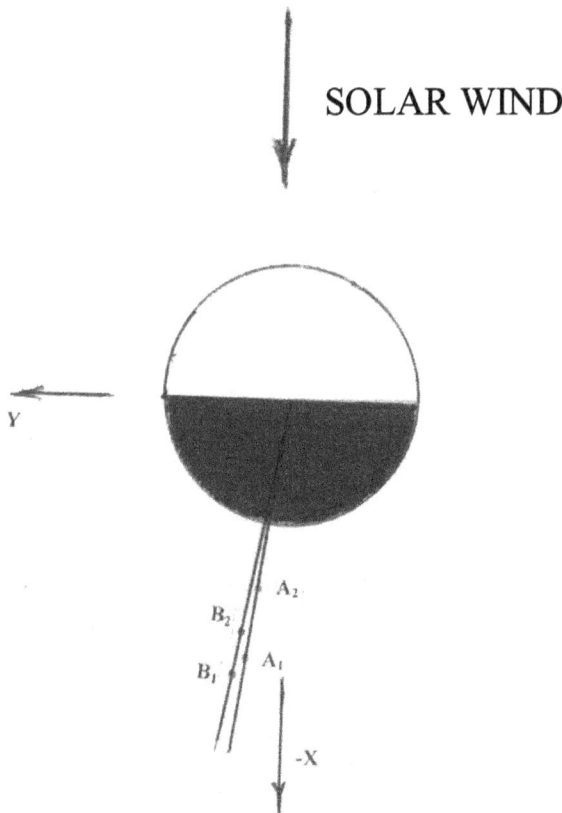

Figure 8.
Projected orientation of VEX orbits on the ecliptic plane XY with data of vortex structures in **Table 1** *during Sept 252,009 (labeled A) and Sept 262,009 (labeled B). The inbound and the outbound VEX crossings have the subscript 1 and 2 for each orbit. In both cases the traces are not directed along the midnight plane but have been shifted in the +Y direction following the effects of the Magnus force on the rotating Venus ionosphere (from [14, 15]).*

values of the O+ ions are observed in those orbits has now been placed in the XY (ecliptic) plane in **Figure 8** to show that they are also shifted to the +Y direction. That region coincides with that where other features produced by the interaction of the solar wind with the Venus ionosphere are shifted in that direction (trans-terminator plasma flow, [16]; ionospheric polar channels, [15]).

Since polar plasma channels are mostly distributed near the midnight plane of the Venus wake rather than by its flanks [15], vortex structures should follow them and as shown in the data of **Table 1** maintain small Y values. However, vortex structures should also be subject to the effects of the aberrated direction of the solar wind and also from the fluid dynamic Magnus force both being directed in the +Y direction. **Figure 8** shows the position of the inbound and outbound VEX encounters of vortex structures along the Sept 25, 2009 and also Sept 26, 2009 orbits that trace the wake with a ~ 10° angle from the midnight plane. It should be noted that both crossings of each orbit occur nearly at the same position in the XY plane despite the fact that they are located at different values along the Z-axis. Thus, vortex structures develop as a near planar structure. Different conditions are encountered in the August 24, 2006 orbit where a wide vortex structure extends along all three coordinates. In fact, similar distances are measured between the inbound and the outbound crossings in the X and in the Y axis. As a result, the structure should not be viewed as a feature that mostly extends in the X direction but that equally applies along the Y direction. Thus, vortex structures may turn out to be complicated features distributed along the wake.

5. Summary of results

A basic issue that is ultimately responsible for the fluid dynamic interpretation employed to account for the motion of the plasma particles within vortex structures is related to their acceleration. Rather than solely applying the convective electric field $E = V \times B$ of the solar wind along their trajectories (V and B being its velocity and magnetic field intensity), it is remarkable that other sources are required to account for the complicated features that are measured. In particular, slingshot trajectories applied to the ionospheric plasma by the magnetic polar regions of Venus and Mars are not sufficient to explain the intricate configuration that is produced from the particle motion and that gives place to the complex vortex flow configuration of their velocity vectors indicated in the left panel of **Figure 1**. A more internal contact between the solar wind and the planetary ions is necessary to deflect their streamlines in a manner that the projection of the velocity vectors on the YZ plane accounts for the peculiar aspect of a vortex flow.

At the same time, while the convective electric field of the solar wind is useful to describe differences in the density and speed of the accelerated planetary ions between the hemispheres where it has a different direction away or toward the planet [12], it is not sufficient to justify the generation of a sharp plasma boundary as that reproduced in **Figure 3** from the Venera measurements. Charge exchange activity between the solar wind hydrogen ions and heavy planetary particles is not satisfactory because it is unrelated to the notable temperature increases reported in those measurements. Instead, concepts based on a fluid dynamic approach rely on arguments that seem to be more accessible to a different acceleration process by invoking wave-particle interactions as the origin of the manner in which both particle populations share their properties. Such condition is expected from the oscillations and fluctuations in the direction and magnitude of the magnetic field measured around the wake [17, 18] and serves to produce the transfer of statistical properties among both plasma populations.

In terms of that description it is of interest to note that it has been useful to account for the following main aspects discussed in this report and that are related to: (i) There is a correlation indicated in **Figure 7** between the width Δ and the extent δ of the vortex structures along the Z axis and the X axis, with $\Delta > \delta$ values implied from the data of **Table 1**; (ii) as noted in **Figure 4,** vortex structures are measured closer to Venus near solar minimum conditions by 2009; (iii) a notable property in the distribution of vortex structures in the Venus wake is the tendency of their width to become smaller with increasing downstream distance from Venus as can be inferred from the position of their segments in **Figure 5**. That difference implies that the thickness of the vortex region decreases along the wake and thus is reminiscent of a corkscrew flow in fluid dynamics represented in **Figure 6**; (iv) an important consequence in the shape of that region is that mass flow conservation across the vortex structure implies larger speed values of particles that move along the wake (particles with larger speeds should be detected far downstream from Venus); (v) as noted above in **Figure 8,** planetary O+ ion fluxes can also be significantly shifted along the Y-axis in response to effects produced by the aberrated direction of the solar wind and the Magnus force on the motion of planetary O+ ions that are dragged by the solar wind.

Acknowledgements

We wish to thank Gilberto A. Casillas for technical work provided. Financial support was available from the INAM-IN108814-3 Project.

Author details

Hector Pérez-de-Tejada[1*] and Rickard Lundin[2]

1 Institute of Geophysics, UNAM, Mexico

2 Swedish Space Research Institute, Kiruna, Sweden

*Address all correspondence to: hectorperezdetejada@gmail.com

IntechOpen

References

[1] Lundin R et al. Ion flow and momentum transfer in the Venus plasma environment. Icarus. 2011;**215**:7

[2] Lundin R et al. A large scale vortex in the Venus plasma tail and its fluid dynamic interpretation. Geophysical Research Letters. 2013;**40**(7):273

[3] Pérez-de-Tejada H et al. Plasma vortex in the Venus wake. Eos. 1982;**63**(18):368

[4] Pérez-de-Tejada H, Lundin R, Intriligator D. Plasma vortices in planetary wakes, Chapter 13. In: Olmo G, editor. Open Questions in Cosmology. Croatia: INTECH Pub; 2012

[5] Pérez-de-Tejada H et al. Solar wind driven plasma fluxes from the Venus ionosphere. Journal of Geophysical Research. 2013;**118**:1-10. DOI: 10.1002/2013JA019029

[6] Ruhunusiri S et al. MAVEN observations of partially developed Kelvin-Helmholtz vortices at Mars. Geophysical Research Letters. 2016;**43**:4763-4773. DOI: 10.1002/2016GL068926

[7] Pérez-de-Tejada H et al. Vortex structure in the plasma flow channels of the Venus wake, Chapter 1. In: Pérez-de-Tejada H, editor. Vortex Structures in Fluid Dynamic Problems. Croatia: INTECH Pub; 2017. DOI: 10.5772/66762

[8] Pérez-de-Tejada H, Lundin R. Solar cycle variations in the position of vortex structures in the Venus wake, Chapter 3. In: Bevelacqua J, editor. Solar System Planets and Exoplanets. Croatia: INTECH-OPEN Pub; 2021. DOI: 10.5772/96710

[9] Acuña M et al. Mars observer magnetic field investigation. Journal of Geophysical Research. 1992;**97**(E5):7799

[10] Romanov SA et al. Interaction of the solar wind with Venus. Cosmic Research. 1979;**16**:603 (Fig 5)

[11] Verigin M et al. Plasma near Venus from the Venera 9 and 10 wide angle analyzer data. Journal of Geophysical Research. 1978;**83**:3721

[12] Dubinin E et al. Plasma in the near Venus tail: VEX observations. Journal of Geophysical Research. 2013;**118**(12):7624

[13] Miller K, Whitten R. Ion dynamics in the Venus ionosphere. Space Science Reviewes. 1991;**55**:165

[14] Pérez-de-Tejada H. Magnus force in the Venus ionosphere. Journal of Geophysical Research. 2006;**111**(A 11). DOI: 10.1029/2005/A01/554

[15] Pérez-de-Tejada H et al. Measurement of plasma channels in the Venus wake. Icarus. 2019;**321**:1026-1037

[16] Phillips J, McComas D. The magnetosheath and magnetotail of Venus. Space Sciences Reviewes. 1991;**55**:1, 1991 (Fig. 35)

[17] Bridge A et al. Plasma and magnetic fields observed near Venus. Science. 1967;**158**:1669-1673 (see Fig. 2)

[18] Vörös Z et al. Intermittent turbulence, noisy fluctuations and wavy structures in the Venusian magnetosheath and wake. Journal of Geophysical Research. 2008;**113**:ED0B21. DOI: 1029/2008JE003159

Chapter 8

Vortex Dynamics in Dusty Plasma Flow Past a Dust Void

Yoshiko Bailung and Heremba Bailung

Abstract

The beauty in the formation of vortices during flow around obstacles in fluid mechanics has fascinated mankind since ages. To beat the curiosity behind such an interesting phenomenon, researchers have been constantly investigating the underlying physics and its application in various areas of science. Examining the behavior of the flow and pattern formations behind an obstacle renders a suitable platform to realize the transition from laminar to turbulence. A dusty plasma system comprising of micron-sized particles acts as a unique and versatile medium to investigate such flow behavior at the most kinetic level. In this perspective, this chapter provides a brief discussion on the fundamentals of dusty plasma and its characteristics. Adding to this, a discussion on the generation of a dusty plasma medium is provided. Then, a unique model of inducing a dusty plasma flow past an obstacle at different velocities, producing counter-rotating symmetric vortices, is discussed. The obstacle in the experiment is a dust void, which is a static structure in a dusty plasma medium. Its generation mechanism is also discussed in the chapter.

Keywords: vortices, vorticity, fluid flow, Reynolds number, plasma, dusty plasma, obstacle, dust void, viscosity

1. Introduction

Vortices are common in fluid motion that originates due to the rotation of fluid elements. They occur widely and extensively in a broad range of physical systems from the earth's surface to interstellar space. A few examples include spiral galaxies in the universe, red spots of Jupiter, tornadoes, hurricanes, airplane trailing vortices, swirling flows in turbines and in different industrial facilities, vortex rings formed by the firing of certain artillery or in the mushroom cloud resulting from a nuclear explosion. The physical quantity that characterizes the rotation of fluid elements is the vorticity $\omega = \nabla \times u$ where u is the fluid velocity. Qualitatively, it can be said that in the region of vortex formation, the vorticity concentration is high compared with its surrounding fluid elements. Vortices formed behind obstacles to a fluid flow are also an interesting observation in various aspects of daily life. Study on the fluid flow around obstacles dates back to the fifteenth century when Leonardo da Vinci drew some sketches of vortex formation behind obstacles in flowing fluids. It has been an interesting and challenging problem in fluid mechanics and is of basic importance in several areas such as the study of aircraft designing, oceanography, atmospheric dynamics, engineering, human blood circulation [1–4]. Analyzing the behavior of flow around such obstacles also provides a medium to study the physical mechanism of transition from laminar to turbulent flow. If a stationary solid boundary lies in the path of a fluid flow, the fluid stops moving on that boundary. Thereby, a boundary layer is formed and its separation from the solid boundary generates various free shear layers that curl into concentrated vortices. These vortices then evolve, interact, become unstable and detach to turbulence. The dynamics of fluids is very diverse and the detail characteristics of transition to turbulence are quite complicated, which also differ from flow to flow. Such understandings can only be realized by experiments and computational models. However, there are a few unifying themes in the theory and a few routes to turbulence that are shared by many flows. One such theme is that when the Reynolds number (the parameter measuring the speed of a class of similar flows with steady configuration) increases, the temporal and spatial complexity of the flows increases eventually leading to turbulence. At a low Reynolds number, a pair of counter-rotating vortices forms behind the cylinder. As the Reynolds number increases, the vortices become unstable and gradually evolve into a von Karman vortex street [5–9]. The topic of flow past an obstacle is of utmost importance from the experimental point of view also. Its understanding is applicable in the stability of submerged structures, vortex-induced vibrations, etc. [10].

Vortices have been extensively studied and explored in the liquid state of matter. However, scientists have also extended their research to study the formation and behavior of vortices in the fourth state of matter, the plasma. Measurements done in space have shown that plasma vortices appear in the earth's magnetosphere as well as along the Venus wake. On both planets, the solar wind encounters different obstacles. For earth, it is the earth's magnetic field and for Venus, the interaction takes place with the ionized components in the upper layer of the planet's atmosphere. Plasma vortices in earth-based laboratories have also been studied theoretically and experimentally [11–15]. Plasma is said to cover more than 99% of the matter found in the universe and dust particles are the unavoidable, omnipresent ingredients in it. Hence, in most cases, plasma and dust particles exist together, and these particles are massive (billion times heavier than the protons). Their size ranges from tens of nanometer to as large as hundreds of microns. Foreign particles in the plasma environment get charged up by the inflow of electrons and ions present in the plasma. The presence of these charged and massive particles increases the complexity of the plasma environment, and hence, this class of plasma has been named as 'complex plasmas' or 'dusty plasmas'. They involve in a rich variety of physical and chemical processes and are thus investigated as a model system for various dynamical processes [16, 17]. Phase transition is an important and characteristic feature in dusty plasmas, due to which it is considered as a versatile medium to study all the three different phases (solid, liquid and gas) in just a single phase. They also behave as many particle interacting systems and provide a unique platform to study various organized collective effects prevalent in fluids, clusters, crystals, etc., in greater spatial and temporal resolution. With the help of laser light scattering, it is possible to visualize the micrometer or nanometer-sized dust particles through proper illumination. This allows to study the various phenomena in dusty plasma in greater spatial and temporal resolution since they appear in a slower time scale owing to their heavier mass [18, 19]. Along with a variety of dynamic phenomena that includes waves, shocks, solitons, etc., dusty plasma medium also supports the formation of vortices. Self-generated vortices have been observed in many dusty plasma experiments, which have been dealt with significant attention. The main cause of such vortex formation is the nonzero curl of the various forces acting on the electrically charged dust particles that are commonly found in radio-frequency (RF) discharges, microgravity conditions and subsonic dusty plasma flow with low Reynolds numbers [20, 21]. The nonzero component of the curl induces a rotational motion to the charged dust particles, which leads to the formation of the vortices. Depending on the different plasma production mechanisms and dust levitation (floating of dust particles in the plasma medium), the causes of the rotation of dust particles vary accordingly. Most importantly, the problem of fluid flow around obstacles can be investigated at the most elementary individual particle level in dusty plasmas. The existence of a liquid phase of dusty plasmas provides us the suitable conditions for the study. However, the obstacles used for such study in dusty plasmas are different from the solid obstacles in the hydrodynamic fluid medium.

In this chapter, we will concentrate mainly on dusty plasmas, their characteristics and a model system to study fluid flow around an obstacle at the particle level. After the introductory portion in Section 1, the fundamentals of dusty plasma are discussed in Section 2. In Section 3, the production of a dusty plasma medium by RF discharge will be discussed. Then in Section 4, we will discuss about the type and behavior of the obstacle which is used in dusty plasma flows. In Section 5, we will discuss the dusty plasma flow and the pattern formation behind the obstacle. The final section then summarizes the chapter as a whole.

2. Fundamentals of dusty plasma

2.1 Dusty plasma

First, let us start with a very brief explanation of plasma! Basically, plasma is an assembly of a nearly equal number of electrons and ions, and the charge neutrality is sustained on a macroscopic scale. In the absence of any external disturbance, that is, under equilibrium conditions, the resulting total electric charge is zero. The microscopic space-charge fields cancel out inside the plasma and the net charge over a macroscopic region vanishes totally. The quasi-neutrality condition at equilibrium is given by,

$$n_e \approx n_i \tag{1}$$

where n_e and n_i are the electron and ion densities, respectively.

The 'plasma' state of matter differs from ordinary fluids and solids by its natural property of exhibiting collective behavior. These collective effects result in the occurrence of various physical phenomena in the plasma, resulting in the long-range of electromagnetic forces among the charged particles. The very first example of plasma that is obvious to refer is the Sun, the source of existence of life. The protective layer to the earth's atmosphere, known as the ionosphere, also remains in the form of ionized particles, that is plasma. Moreover, natural plasmas exist in interstellar space, stars, intergalactic space, galaxies, etc. On earth, the common form of natural plasma is lightning, fire and the amazing Aurora Borealis. Artificial plasmas are generated by applying electric or magnetic fields through a gas at low pressures. These are commonly found in street lights, neon lights, etc. Neon light is a gas discharge light, which is actually a sealed glass tube with metal electrodes at both the ends of the tube and filled with one or several gases at low pressure.

As already mentioned before, dust particles in space as well as in earth's atmosphere, are unavoidable. These particles in plasma form a new field, that is dusty plasma or complex plasma. Dusty plasma is defined as a normal electron-ion plasma with charged dust components added to it. Naturally, dust grains are metallic, conducting, or made of ice particulates. Until and unless these are manufactured in laboratories, their shape and size vary. Depending on the surrounding plasma environments (due to the inflow of electrons and ions), dust particulates are either negatively or positively charged. These charged particles as a whole affect the plasma and result in collective and unusual behavior. When observed from afar, dust particles can be considered as point charges. As they are charged by the plasma species (electrons and ions), the charge neutrality condition is now modified, which is given by,

$$Q_d n_{d0} + e n_{eo} = e n_{io}$$
(2)

where n_{e0}, n_{i0} and n_{d0} are the equilibrium densities of electrons, ions and dust, respectively, 'e' is the magnitude of electron charge, $Q_d = e Z_d$ is the charge on the dust's surface and Z_d is the dust charge number. It is important to highlight that the charge of the particles depends significantly on the plasma parameters. And the basic physics of the dusty plasma medium entirely rests on the $Q_d n_{d0}$ term of the charge neutrality condition.

Plasma possesses the fundamental property of shielding any external potential by forming a space charge around it. This particular property provides a measure of the distance over which the influence of the electric field of a charged particle (dust particle in our case) is experienced by other particles (electrons and ions) inside the plasma. Typically, this length is known as the dust Debye length λ_D, within which the dust particles can rearrange themselves to shield all the existing electrostatic fields. The negatively charged heavier dust particles are assumed to form a uniform background and the electrons and ions, which are assumed to be in thermal equilibrium, simply obey the Boltzmann distribution. The dust Debye length is given by,

$$\lambda_D = \frac{\lambda_{De}\lambda_{Di}}{\sqrt{\left(\lambda_{De}^2 + \lambda_{Di}^2\right)}}$$
(3)

where λ_{De} and λ_{Di} are electron and ion Debye lengths, respectively. These are expressed as,

$$\lambda_{De} = \sqrt{\frac{k_B T_e}{4\pi n_{eo} e^2}}$$
(4)

$$\lambda_{Di} = \sqrt{\frac{k_B T_i}{4\pi n_{io} e^2}}$$
(5)

$T_{e,i}$ represents the electron and ion temperatures, respectively, $n_{eo,io}$ are the electron and ion densities, respectively, and k_B is the Botlzmann constant.

A pictorial representation of a dusty plasma medium is shown in **Figure 1**.

In a dusty plasma medium, the charged particles interact with each other *via* the electrostatic Coulomb force. However, due to the inherent shielding property of the plasma electrons and ions, the charged particles are shielded and hence, the interaction energy among them is known as Screened Coulomb or Yukawa potential energy. Consider two dust particles having the same charge Q_d and separated by a distance 'a'. The screened Coulomb potential energy is given by,

$$P.E = \frac{Q_d^2}{4\pi\epsilon_0 a} e^{-\kappa}$$
(6)

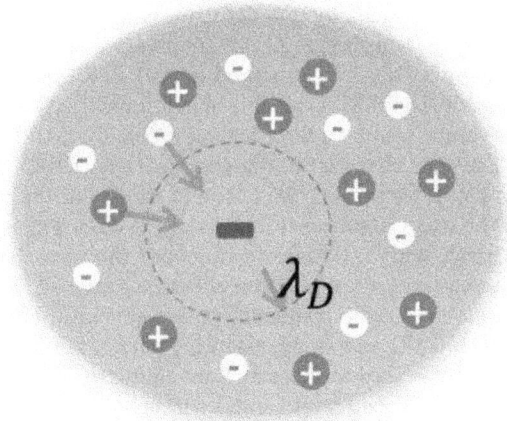

Figure 1.
Schematic of a dusty plasma medium. The pink-shaded portion is the plasma medium. The green ball is the dust particle that is negatively charged. λ_D is the dust Debye length.

where $\kappa = \frac{a}{\lambda_D}$ is the screening strength. The dust thermal energy is given by,

$$K.E = k_B T_D \tag{7}$$

where T_D is the dust temperature. The ratio of the P.E to the K.E is termed as the Coulomb Coupling parameter, given by

$$\Gamma = \frac{Q_d^2}{4\pi\epsilon_0 a k_B T_D} e^{-\kappa} \tag{8}$$

Depending on the coupling parameter, a dusty plasma system remains in a weakly coupled state or a strongly coupled one. When $\Gamma < 1$, the thermal energy of the dust particles is greater than the potential energy of the system and the system is said to be weakly coupled. On the other hand, when the potential energy exceeds the thermal energy, that is $\Gamma > 1$ the system becomes strongly coupled. So, from Eq. (8), we can see that dust charge, screening parameter and the dust temperature play an important role in determining the system's coupling state. As Γ exceeds a critical value Γ_c, called the critical coupling parameter, a dusty plasma system attains a crystalline state. However, this critical value for crystallization is dependent on the screening parameter [22]. For $1 < \Gamma < \Gamma_c$, the system remains in a fluid (liquid or gas) state.

Thus, we see that by adjusting the dusty plasma parameters, we can obtain a fluid state of the dusty plasma medium experimentally. This provides us a unique model to study vortex formation behind an obstacle in the particle most level.

3. Production of a dusty plasma medium

Laboratory dusty plasmas differ from space and astrophysical dusty plasmas in a significant manner. The discharges done in the laboratory have geometrical boundaries. The composition, structure, conductivity, temperature, etc., of these geometries affect the formation and transport of the dust grains. Also, the external circuit, which produces and sustains the dusty plasma, imposes boundary conditions on the dusty discharge, which vary spatially as well as temporally. Dusty plasmas in the laboratory are generally produced by two main discharge techniques—direct current (DC) discharge and RF discharge. In this chapter, we will mainly focus on the production technique by RF discharge method in a DUPLEX device.

As the name suggests, DUPLEX is an abbreviation for Dusty Plasma Experimental Chamber. It comprises of a cylindrical glass chamber, 100 cm in length and 15 cm in diameter. The glass chamber configuration of the DUPLEX device provides a suitable and great access for optical diagnostics. One end of the cylindrical chamber is connected to the vacuum pump systems and the other end is closed by a stainless steel (SS) flange with Teflon O-ring between the glass chamber and the SS flange. On this closed end, there are ports for pressure gauge fitting, probe insertion and electrical connections. A radio frequency power generator (frequency: 13.56 MHz, power: 0–300 W) and an RF matching network are used for the plasma discharge. The RF antennas used in this setup are aluminum strips of 2.5 cm width and 20 cm length typically placed on the outer surface of the glass chamber. A schematic of the setup is shown in **Figure 2**. This strip acts as the live electrodes.

Initially, the chamber pressure is reduced to a value of about $\sim 10^{-3}$ mbar with the help of a rotary pump. Argon is used as the discharge gas, by injecting which the desired chamber pressure can be maintained. A grounded base plate is also inserted into the chamber (about $\sim$30 cm length, 14.5 cm width and 0.2 cm thickness), which acts as the grounded electrode and the region above it is selected as the experimental region. Applying a radiofrequency power (13.56 MHz and 5 W) between the aluminum strips (working as live electrodes) and the grounded base plate, a capacitively coupled RF discharge plasma is produced. Due to the application of the RF field, initially, the stray electrons inside the chamber get energized and in turn ionize the gas molecules present in the chamber. The aluminum strips used as live electrode outside give the flexibility to change the electrode position whenever required. Also, it facilitates in forming a uniform plasma over an extensive area of the grounded plate, that is the experimental region. The plasma parameters can be varied manually by tuning the discharge conditions, *viz*. RF power and neutral pressure.

Dust particles used in the experiment are gold-coated silica dust particles ($\sim$ 5 micron diameter). These are initially put inside a buzzer that is fitted to the grounded base plate. After the production of the plasma, a direct current (DC) voltage of $\sim$ (6–12)V is applied to the buzzer, which ejects the dust particles from it

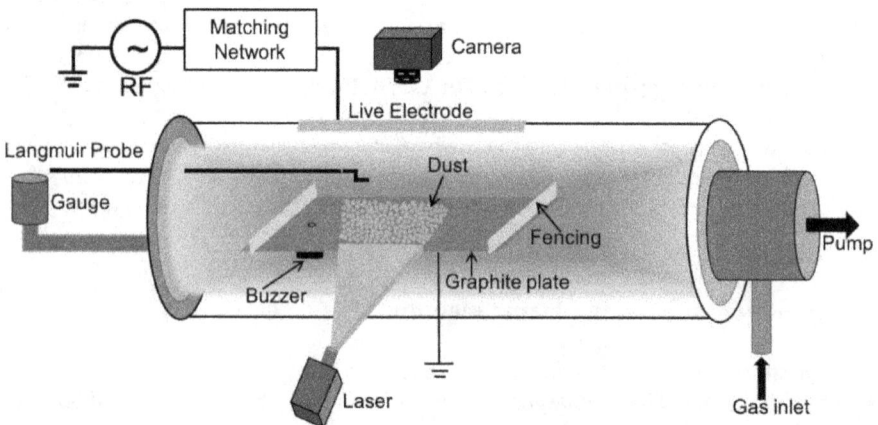

Figure 2.
Schematic of a DUPLEX setup. The pink-shaded portion is the argon plasma.

through a hole. When these dust particles enter into the plasma environment, electrons and ions flow towards it and charge up the particles. In the laboratory, the dust particles are usually negatively charged as the electrons are lighter and highly mobile than the ions. These negatively charged dust particles are acted upon by two forces mainly, the upward electric field force (Q_dE) due to the sheath electric field (E) of the grounded base plate and the downward gravitational force (m_dg). Dust particles levitate at the position where these two forces exactly balance. The dust particles are illuminated by laser light scattering, and the dust dynamics are recorded in high-speed cameras. **Figure 3** shows the levitation of dust particles in a plasma medium. Above the dust layer, the purple color signifies argon discharge plasma. The dark region below the dust layer and above the plate is the sheath (where ionization does not take place) where a strong electric field (E) is present. The charged dust particles levitate at the interface region (~ 0.8 cm above the plate) between plasma and the sheath where the force balance occurs. This is shown by a dashed line.

4. Obstacle in dusty plasma flows

The obstacle used in dusty plasma flow experiments is actually a dust void. A void is a dust-free region, which is encountered spontaneously in certain experimental conditions or can be produced externally also [23–28]. In the past couple of decades, there have been a few studies on the interaction of a dusty plasma medium with dust voids. In 2004, Morfill et al. studied a laminar flow of liquid dusty plasma with a velocity ~ 0.8 cms^{-1} around a spontaneously generated lentil-shaped void [29]. They observed the formation of a wake behind the void that is separated from the laminar flow region by a mixing layer. The flow also exhibited stable vortex flows adjacent to the boundary of the mixing layer. Another study was made in 2012

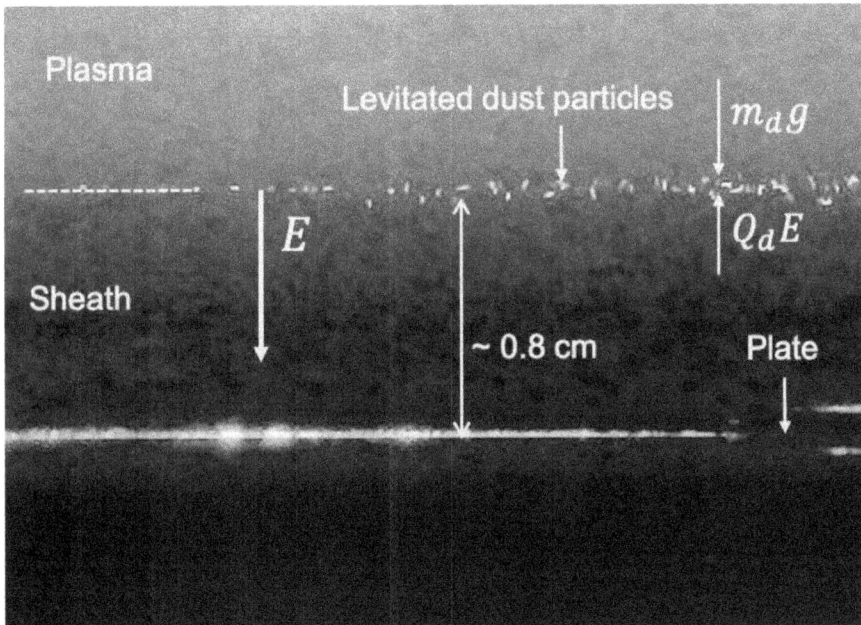

Figure 3.
Photograph of a dust layer levitation in plasma.

by Saitou et al. where they externally placed a thin conducting wire of 0.2 mm diameter and 2.5 cm length. They made the dust particles flow with velocity in the range $\sim (5–15)$ cms^{-1} but did not observe any vortex formation behind the obstacle. What they observed was a bow shock in front of it [30]. In the very next year itself (2013), Meyer et al. also did a similar experiment with a different configuration and dust flow mechanism (velocity $\sim 10–25$ cms^{-1}) than Saitou's [21]. They produced a dust void by placing a 0.5-mm-diameter cylindrical wire transverse to the flow. They too observed a bow shock and a tear-shaped wake in front and behind the obstacle, respectively. Moreover, Charan et al. in 2016 did a molecular dynamics simulation study where they used a square obstacle and observed von Karman vortex street at low Reynolds number (i.e. low velocity) compared with normal hydrodynamic fluids [31]. Then in 2018, Jaiswal et al. investigated dust flow towards a spherical obstacle over a range of flow velocities $\sim (4–15)$ cms^{-1} and different obstacle biases [32]. The spherical obstacle also generated a dust-free area in its vicinity. They too observed bow shock formation in front of the obstacle but no vortex formation behind it. In 2020, Bailung et al. also investigated the study of dust flow around a dust void with a unique flow mechanism (dust flow velocity $\sim 3–10$ cms^{-1}) in a DUPLEX setup [33]. Dust particles are allowed to flow towards an already existing stationary dust layer. They could observe the formation of a counter-rotating pair of vortices behind the obstacle in a particularly narrow range of velocity $\sim (4–7)$ cms^{-1}. Above and below this range, their vortices are not observed. Due to the interplay between these two forces, a circular void is generated around the pin. At the void boundary, these two forces equate with each other.

In the next section, we will study the results of Bailung et al. in detail, but before that let us understand the mechanism of the formation of dust void due to the insertion of an external cylindrical wire. A cylindrical pin inside the plasma attains a negative potential for the plasma and a sheath is formed in its vicinity. Due to the negative potential of the pin, ions try to drift towards it giving rise to a force on the dust particles named as ion drag force. This force is directed radially inward with the pin as the centre. Also, the negatively charged dust particles experience a repulsive electrostatic force from the pin which is directed radially outward. The interplay between these two forces generates a circular void around the pin. At the void boundary, these two forces equate with each other. A typical configuration of pin insertion through the grounded plate of a DUPLEX chamber is shown in **Figure 4**. The pin is externally connected to a DC bias voltage. By varying the bias voltage, the size of the void can be altered according to experimental requirements. Typically, at a RF power of 5 W and chamber pressure ~ 0.02 mbar, the diameter of the dust void in floating condition (i.e. no external bias) is ~ 1.7 cm. A typical example of a dust void is shown in **Figure 5**. However, unlike the solid obstacles in the case of hydrodynamic fluids, the dust void is not a rigid kind of obstacle. As already seen, the boundary of the void is maintained by a delicate force balance

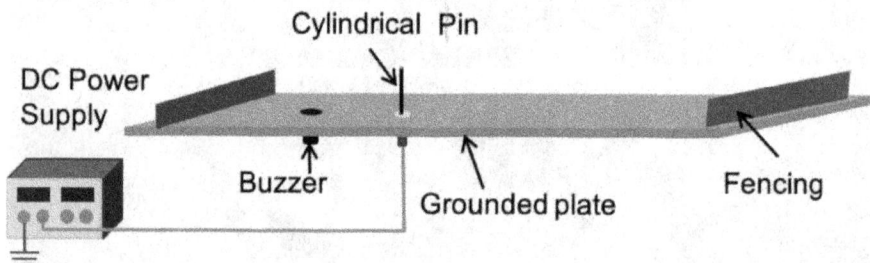

Figure 4.
A typical configuration for insertion of a pin through a grounded base plate in DUPLEX chamber.

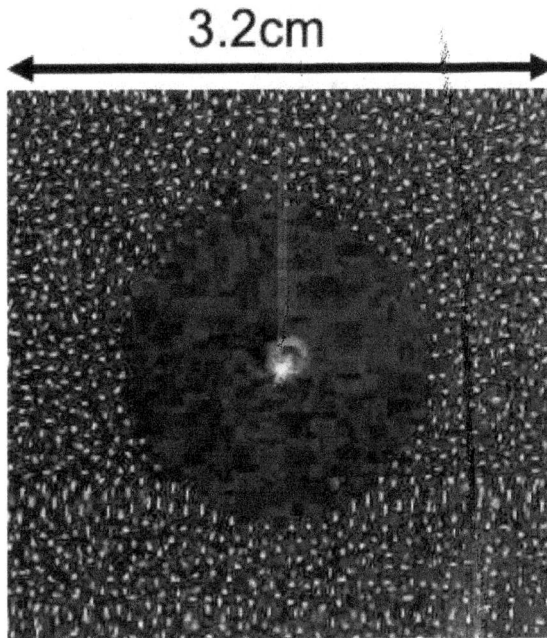

Figure 5.
Snapshot of a dust void formed in DUPLEX chamber. The bright spot in the Centre is the reflection of laser light from the cylindrical pin. The photograph is taken from the top of the chamber.

between the outward electrostatic force and inward ion drag force. An incoming dust flow, depending on the velocity of the flow, would cross the void boundary and penetrate into the void.

5. Vortices in the wake of a dust void

Due to the non-rigidity of the dust void boundary, the behavior of the flow near the obstacle is somewhat different than conditions of hydrodynamic fluid with a rigid obstacle. Despite this difference, the transition from laminar to turbulence is observed in the wake of the obstacle in the case of dusty plasma flow also. As the flow approaches the void boundary, the middle section of the flow slightly penetrates into the void region and slips through the void boundary layer on both sides. The trajectory of the flow (in the mid-section) is deflected in front of the void due to the repulsive force exerted by the sheath electric field of the void and then flows downstream surrounding the void. The curved dust flow again meets behind the void and continues with the flow. As observed by Bailung et al. at a very narrow range of velocity $\sim$ (4–7) cms^{-1}, a counter-rotating vortex pair is seen to appear. Below and above this range, the dust particles do not form any vortices. A typical example of three different conditions is shown in **Figure 6**.

In each of the images, dust particles flow from right to left shown by dashed arrows. The top image (a) depicts a flow with dust flow velocity $\sim$ 3.5 cms^{-1} and the snapshot is at time t = 1370 ms from a reference time (t = 0, when dust flow reaches the right edge of the images). The middle image (b) shows dust fluid flow velocity $\sim$ 4.5 cms^{-1} at t = 1033 ms showing a vortex pair formation behind the void. The vortices are shown by the two arrow marks. It is observed that vortices are not formed for larger flow velocity $\sim$ 8 cms^{-1} (image (c)). For such high

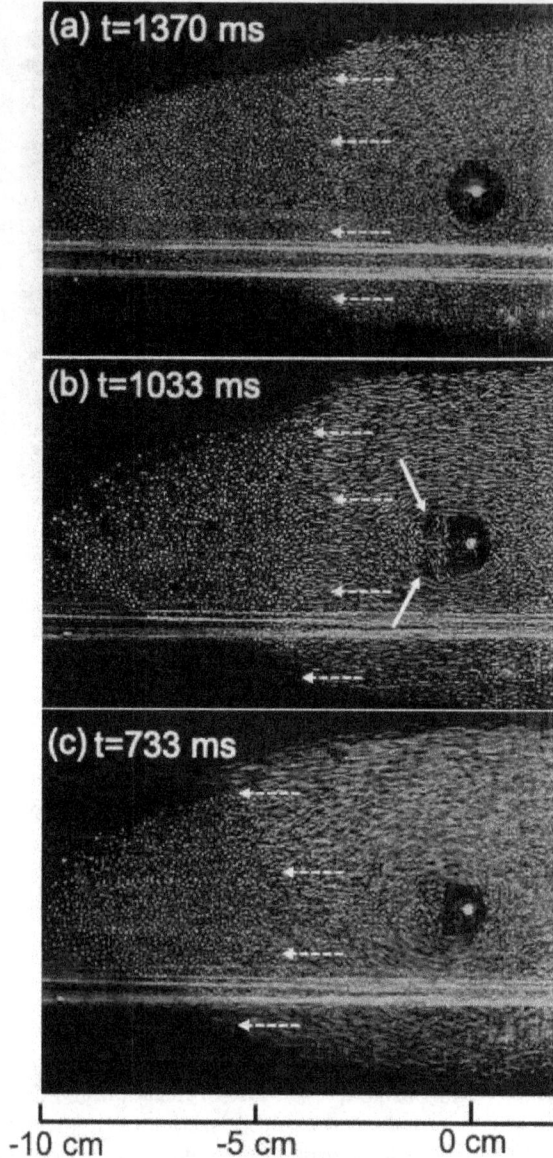

Figure 6.
Typical snapshots showing structures formed behind the void at (a) 3.5cms^{-1}, (b) 4.5cms^{-1}, (c) 8 cm^{-1}.

velocities, flow trajectories behind the void are elongated and dynamics in the wake is rather complex due to cross-flow at high speed. The bright illuminated point at the centre of each image is the reflection of laser light from the pin. Two horizontal lines that appear in all the images are due to laser reflection from the wall of the glass chamber. It is noted that dust flow with unsteady laminar velocity, which is $(4-7)$cms^{-1}, and optimum dust density in the experimental region above the grounded plate is required to generate the vortex behind the void.

For a better understanding, a pictorial representation showing the dusty plasma streamlines around dust void at three different velocities are shown in **Figure 7(a)–**
(c) of **Figure 7** corresponds to the observation shown in (a), (b) and (c) of **Figure 6**.

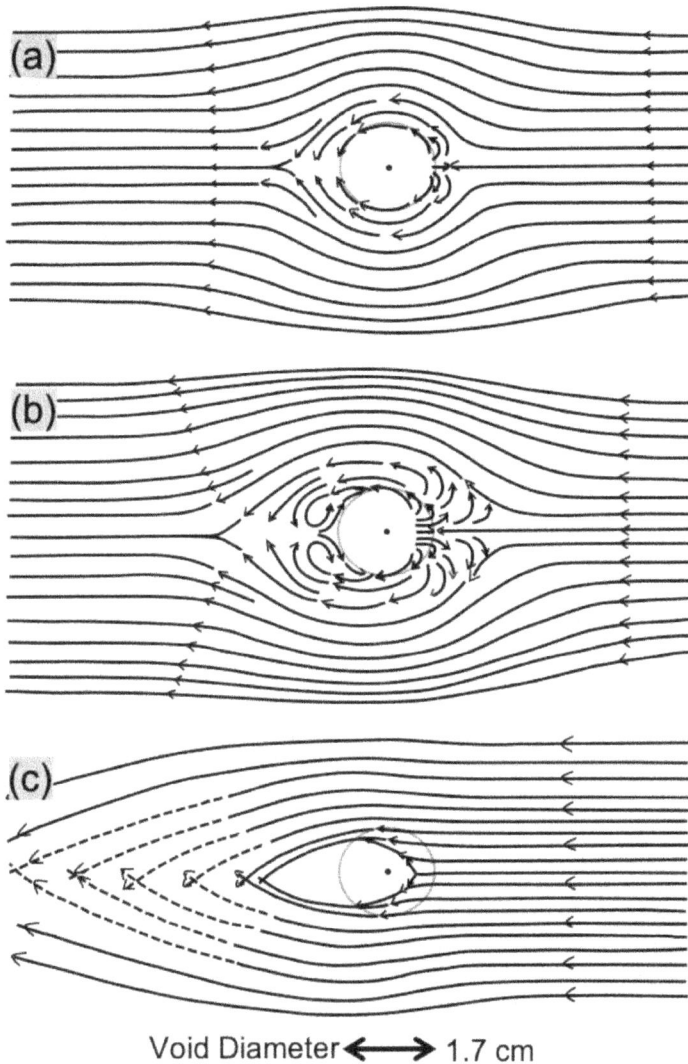

Figure 7.
A pictorial illustration of the dusty plasma flow interaction with the dust void at different flow velocities. (a) Laminar flow, (b) unsteady laminar flow with filamentary vortex-type structure in the upstream and vortex pair in the downstream and (c) turbulent flow.

At a lower dust flow velocity, the void in the upstream is slightly compressed and trajectories of the streamlines flowing close to the void (boundary layer) curl behind the void. However, no structure formation in the wake appears here. Dust particles, after meeting behind the void, just continue with the flow smoothly. For critical flow speed (b), flow dynamics in the upstream void boundary is quite different. Streamlines that hit perpendicularly at the void flow some distance into the void region. They reconstruct the boundary during the flow and get ejected backward making the streamline bifurcation to occur much ahead of the void boundary. The curved streamlines, which are ejected backward, again flow along with the incoming dust flow close to the boundary layer. This critical reorientation in the front of the void generates a suitable condition for the formation of the vortex pair behind the void. Particles get slowed down in this region and these slower particles flow close to the boundary layer around the void and contribute in the formation of the vortex pair. At higher velocities (c), that is above the critical range for vortex formation, all the particles that hit the upstream void boundary are flushed away by the flow along with it. The streamlines intersect and crossover at a distance far behind the void and there is no formation of any stable structure. It is well known that in hydrodynamic fluids, at much higher velocities, vortex streets are observed. However, here such streets are not observed to form. This may be due to the restriction of the experimental geometry. The transition from laminar to turbulence is well known in fluid dynamics. But studying it in dusty plasma provides the chance to observe the individual particle-level trajectory. In turn, the dynamics can be studied in greater detail.

To see the dust dynamics in greater detail, let us look at the vortex formation behind the dust void step by step. At the outset of the formation, the slower particles moving along the curved boundary layer interact with the stationary particles behind the void and start to swirl on each side. The flow front then meets in the wake region behind the void (**Figure 8(a)**) and gradually traverses a swirling circular path. This is evident in the dotted arrow marks in (Figure b). After duration of 966 ms from the start of the flow, two counter-rotating vortices complete their formation (Figure c). Only the slower particles flowing close to the boundary layer participate in this swirling motion due to the nonzero curl of the forces. Those particles away from the boundary layer move faster and do not contribute to the swirling. With increasing time and inflow of more particles, the swirling finally grows into a distinct pair of the vortex with an eye in the middle (Figure (d)). As the flow progresses by maintaining a constant inflow of particle flux, the vortex pair sustains till 1167 ms. The one shot of dust flow in the experiment done by Bailung et al. lasted for about 2 sec.

Hence, gradually when the particle flux started decreasing, the vortex pair starts to die out. It is faintly visible till 1233 ms (Figure g). The time for the growth of the vortex pair is ∼200 ms (from the time the particles meet behind the void) and survives for duration of 200 ms (depending on the duration of accelerated dusty plasma fluid flow). Finally, they disappear after 1300 ms. The rotational frequency measured for the vortices is about ∼3 Hz.

It is already mentioned that the advantage of studying vortex dynamics in dusty plasma lies in the fact that particles can be individually tracked. Different particle tracking software and computational models are available, which can

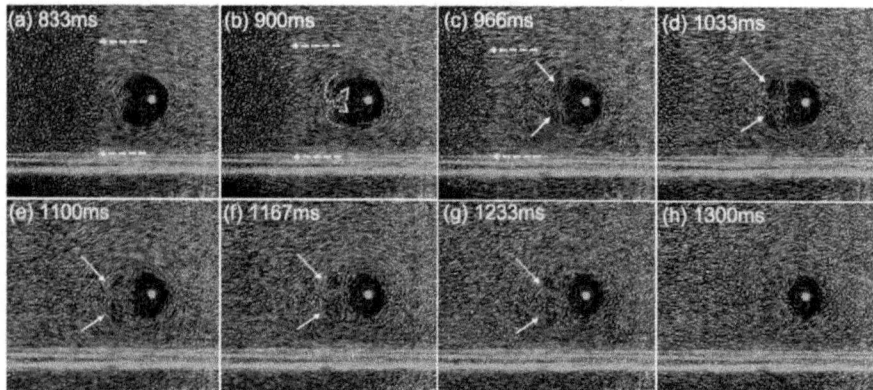

Figure 8.
The parallel arrows depict the direction of the dust flow. (a) when both the oppositely curling flow front meet behind the void. Dotted curve traces in (b) indicate flow trajectories. The arrows in (c) - (g) show the vortex pair. the vortices vanish with time when flow is nearly over (h).

generate the velocity vectors of the trajectory of the particles and hence can give a quantitative interpretation of the experimentally observed results. One such particle tracking platform is OpenPIV (Open Particle Image Velocimetry) in MATLAB [34]. This helps to study the evolution of the vortex pair along with its vorticity. But to perform successful PIV from images, the recorded videos of the dust flow dynamics should have a high-quality resolution and must be in high speed. A PIV analysis performed on a video recorded at 100 frames per second is shown in **Figure 9**.

Each image in the figure is an average PIV result of 10 consecutive image frames. The position of the void and the pin position are drawn by a red-dashed circle and a red dot, respectively. The velocity vectors show the trajectory of the dust particles and the color code gives the value of the vorticity at different times in units of s^{-1}. The slowing down of the particles in front of the void is clearly seen by comparing the velocity vectors' lengths in **Figure 9(a)** and **(b)**. The backflow of the incoming dust particles mentioned earlier (due to repulsive sheath electric field force of the pin) is also observed in (b). The curling of dust particles leading to vortex formation is evident from (c) and (d). The vorticity of the fully formed vortex pair is about $\sim (20\text{--}25)s^{-1}$, which is shown by the color bar in (e) and (f)). This is nearly equal to twice the measured angular frequency. With the decrease of the dust flow influx, the vortex structure deforms (vorticity $\sim 15\ s^{-1}$) and breaks away into smaller vortices (vorticity $\sim 10s^{-1}$) as seen from (g) to (i). Vortices finally disappear in (j), evident from the vorticity value which almost tends to 0.

Reynolds number is the characteristic parameter that helps to predict flow patterns. It is the ratio of the inertial forces to the viscous forces and is given by,

$$Re = \frac{\rho v_d L}{\eta} \tag{9}$$

where ρ is mass density, v_d is dust velocity, L is the obstacle dimension, that is the void diameter and η is the viscosity of the dust fluid.

In case of dusty plasma fluids, the viscosity is estimated from the formula,

$$\eta = \sqrt{3}\hat{\eta} m_d n_d \omega_E a^2 \tag{10}$$

where $\hat{\eta}$ is the normalized shear viscosity, m_d is the dust mass, n_d is the dust density, $\omega_E = (\omega_{pd}/\sqrt{3})$ is the Einstein frequency, ω_{pd} is the dust plasma frequency and a is the interparticle distance. The normalized shear viscosity in dusty plasma fluid is a function of the Coulomb coupling parameter Γ, which has been estimated for a range of coupling parameters in different conditions *via* simulation [35, 36]. For typical plasma parameters of DUPLEX chamber, that is,

$$m_d = 1.7 \times 10^{-13}\ kg$$

$$n_d = 9 \times 10^9 m^{-3}$$

$$\omega_{pd} = 247.5\ s^{-1}$$

$$a = 3 \times 10^{-4}\ m$$

The viscosity is calculated to be 9×10^{-9} Pas.

Thus, the Reynolds number for dust flow velocity $\sim (3\text{--}10)\ cms^{-1}$ is estimated to be lying in the range 50–190. The vortex pair formation appears in a critical range of 60–90.

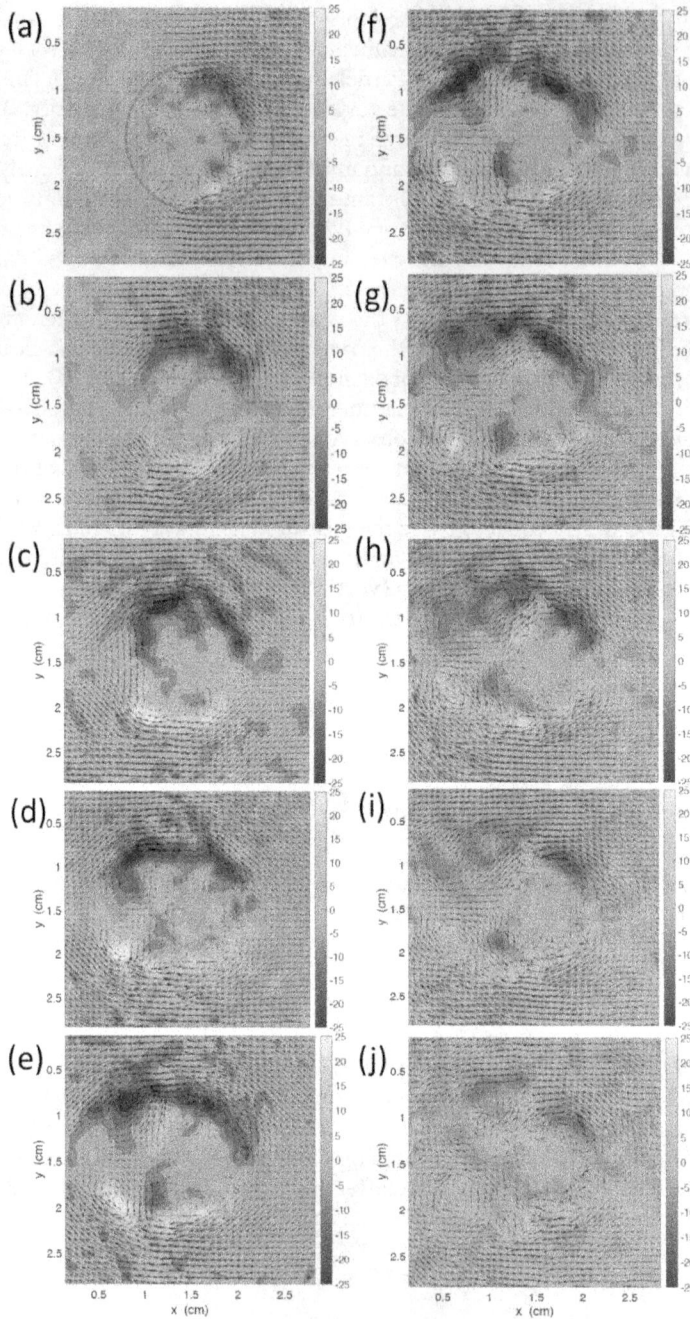

Figure 9.
PIV analysis showing the time evolution of the vortices for a duration of 1 sec. The velocity vectors and vorticity profile drawn from (a) (0.53-0.62) sec (b) (0.63-0.72) sec (c) (0.73-0.82) sec (d) (0.83-0.92) sec (e) (0.93-1.02) sec (f) (1.03-1.12) sec (g) (1.13-1.22) sec (h) (1.23-1.32) sec (i) (1.33-1.42) sec (j) (1.43-1.52) sec are shown. The color bar shows the value of vorticity in 1/s. The dotted circle in (a) shows the original position of the void boundary before the flow and the dot at the center of the circle depicts the pin position.

In the case of hydrodynamic fluid, the range of Reynolds number for vortex formation is 5–40, which is much lower compared with that in dusty plasma fluid. This is because the ratio ρ/η (which is the kinematic viscosity) is one order larger in the case of dusty plasma fluids than that in hydrodynamic fluids. The estimated kinematic viscosity for dusty plasma fluids is $\sim$0.088 cm^2s^{-1}, whereas the kinematic viscosity for water is $\sim$0.008 cm^2s^{-1}.

6. Conclusion

The study of vortices in the problem of flow past an obstacle is significant as it provides a platform to investigate the transition from laminar to turbulence. Formation of vortices in the wake region behind an obstacle appears in the unsteady laminar regime of flows and has been widely studied in hydrodynamic fluids. However, dusty plasma medium, which is a component of the fourth state of matter, provides a unique stage to study such phenomena at the particle level. A special property of this medium is that it can remain in both fluids (liquid- or gas-like) as well as the crystalline state. By mere adjustment of plasma conditions, the desired state can be obtained. The individual tracking of micron-sized dust particles by methods such as PIV (Particle Image Velocimetry) yields the particle trajectory in form of velocity vector fields. This gives a very clear picture of the behavior of flow near obstacle boundaries. However, the obstacles used in dusty plasma flow experiments differ from those in hydrodynamic fluid experiments in the sense that unlike those in hydrodynamics, the obstacle boundaries in dusty plasma are non-rigid. Any foreign pin or wire inserted into the plasma would possess a negative potential with respect to the plasma. Dust particles in its vicinity are repelled due to electrostatic force and form a dust-free region around it, called the dust void. This dust void, whose boundary is delicately maintained by dusty plasma forces, acts as a non-rigid type of obstacle. Dusty plasma flows also generate counter-rotating vortices in the wake region behind a dust void at a particular range of velocities. Below and above this range, no structure formation is seen to appear. The particle behavior causing the formation of the vortices is better understood by tracking particles in consecutive frames. The estimated Reynolds number value for vortices to appear in the wake of a void in a dusty plasma medium is estimated to lie in the range 60–90. This is quite larger than the Reynolds number range for hydrodynamic fluids which is roughly about 5–40. This higher range in dusty plasma medium is attributed to the higher kinematic viscosity of dusty plasma fluids. However, in dusty plasma experiments, Von Karman vortex streets (observed in the turbulent regime of hydrodynamic fluids) are not yet explored. If such experiments could be successfully performed, then there will be immense scope of understanding turbulence at the particle-most level and with a better perspective. Although to study turbulent dynamics, high-speed cameras with high-quality resolution would be necessary.

Author details

Yoshiko Bailung[1] and Heremba Bailung[2]*

1 Department of Physics, Goalpara College, Goalpara, India

2 Dusty Plasma Laboratory, Physical Sciences Division Institute of Advanced Study in Science and Technology, Guwahati, Assam, India

*Address all correspondence to: hbailung@yahoo.com

IntechOpen

References

[1] Lopez HM, Hulin JP, Auradou H, D'Angelo MV. Deformation of a flexible fiber in a viscous flow past an obstacle. Physics of Fluids. 2015;**27**: 013102:1-12

[2] Saffman PG, Schatzman JC. Stability of a vortex street of finite vortices. Journal of Fluid Mechanics. 1982;**117**: 171-185

[3] Zhang HQ, Fey U, Noack BR, Knig M, Eckelmann H. On the transition of the cylinder wake. Physics of Fluids. 1995;**7**:779-794

[4] Vasconcelos GL, Moura M. Vortex motion around a circular cylinder above a plane. Physics of Fluids. 2017;**29**: 083603:1-8

[5] Votyakov EV, Kassinos SC. On the analogy between streamlined magnetic and solid obstacles. Physics of Fluids. 2009;**21**:097102:1-11

[6] Chenand D, Jirka GH. Experimental study of plane turbulent wakes in a shallow water layer. Fluid Dynamics Research. 1995;**16**:11-41

[7] Schar C, Smith RB. Shallow water flow past topography. Part II: Transition to vortex Shedding. Journal of the Atmospheric Sciences. 1993;**50**: 1401-1412

[8] Balachandar R, Ramachandran S, Tachie MF. Characteristics of shallow turbulent near wakes at low Reynolds number. Journal of Fluids Engineering. 2000;**122**:302-308

[9] Koochesfahani MM. Vortical patterns in the wake of an oscillating airfoil. AIAA Journal. 1989; 27:1200-1205

[10] Bharuthram R, Yu MY. Vortices in an anisotropic plasma. Physics Letters A. 1987;**122**:488-491

[11] Kervalishvili NA. Electron vortices in a nonneutral plasma in crossed E⊥H fields. Physics Letters A. 1991;**157**: 391-394

[12] Kaladze TD, Shukla PK. Self-organization of electromagnetic waves into vortices in a magnetized electron-positron plasma. Astrophysics and Space Science. 1987;**137**:293-296

[13] Siddiqui H, Shah HA, Tsintsadze NL. Effect of trapping on vortices in plasma. Journal of Fusion Energy. 2008;**27**:216-224

[14] Mofiz UA. Electrostatic drift vortices in a hot rotating strongly magnetized electron-positron pulsar plasma. Astrophysics and Space Science. 1992;**196**:101-107

[15] Horton W, Liu J, Meiss JD, Sedlak JE. Solitary vortices in a rotating plasma. Physics of Fluids. 1986;**29**: 1004-1010

[16] Morfill GE, Ivlev AV. Complex plasmas: An interdisciplinary research field. Reviews of Modern Physics. 2009; **81**:1353-1404

[17] Shukla PK. A survey of dusty plasma physics. Physics of Plasmas. 2001;**8**: 1791-1803

[18] Burlaga LF. A heliospheric vortex street. Journal of Geophysical Research. 1990;**95**:4333-4336

[19] Hones EW, Birn J, Bame SJ, Asbridge JR, Paschmann G, Sckopke N, et al. Further determination of the characteristics of magnetospheric plasma vortices with Isee 1 and 2. Journal of Geophysical Research. 1981; **86**:814-820

[20] Schwabe M, Zhdanov S, Rath C, Graves DB, Thomas HM, Morfill GE. Collective effects in vortex movements

in complex plasmas. Physical Review Letters. 2014;**112**:115002:1-5

[21] Meyer JK, Heinrich JR, Kim SH, Merlino RL. Interaction of a biased cylinder with a flowing dusty plasma. Journal of Plasma Physics. 2013;**79**: 677-682

[22] Ichimaru S. Strongly coupled plasmas: High-density classical plasmas and degenerate electron liquids. Reviews of Modern Physics. 1982;**54**: 1017

[23] Praburam G, Goree J. Experimental observation of very low-frequency macroscopic modes in a dusty plasma. Physics of Plasmas. 1996;**3**:1212-1219

[24] Samsonov D, Goree J. Instabilities in a dusty plasma with ion drag and ionization. Physical Review E. 1999;**59**: 1047-1058

[25] Morfill GE, Thomas H, Konopka U, Rothermel H, Zuzic M, Ivlev A, et al. Condensed plasmas under microgravity. Physical Review Letters. 1999;**83**: 1598-1601

[26] Rothermel H, Hagl T, Morfill GE, Thoma MH, Thomas HM. Gravity compensation in complex plasmas by application of a temperature gradient. Physical Review Letters. 2002;**89**: 175001:1-4

[27] Fedoseev AV, Sukhinin GI, Dosbolayev MK, Ramazanov TS. Dust-void formation in a dc glow discharge. Physical Review E. 2015;**92**:023106:1-9

[28] Bailung Y, Deka T, Boruah A, Sharma SK, Pal AR, Chutia J, et al. Characteristics of dust voids in a strongly coupled laboratory dusty plasma. Physics of Plasmas. 2018;**25**: 053705:1-8

[29] Morfill GE, Zuzic MR, Rothermel H, Ivlev AV, Klumov BA, Thomas HM, et al. Highly resolved fluid flows:

"Liquid plasmas" at the kinetic level. Physical Review Letters. 2004;**92**: 175004:1-4

[30] Saitou Y, Nakamura Y, Kamimura T, Ishihara O. Bow shock formation in a complex plasma. Physical Review Letters. 2012;**108**:065004:1-4

[31] Charan H, Ganesh R. Molecular dynamics study of flow past an obstacle in strongly coupled Yukawa liquid. Physics of Plasmas. 2016;**23**:123703:1-7

[32] Jaiswal S, Schwabe M, Sen A, Bandyopadhyay P. Experimental investigation of dynamical structures formed due to a complex plasma flowing past an obstacle. Physics of Plasmas. 2018;**25**:093703:1-10

[33] Bailung Y, Chutia B, Deka T, Boruah A, Sharma SK, Kumar S, et al. Vortex formation in a strongly coupled dusty plasma flow past an obstacle. Physics of Plasmas. 2020;**27**:123702:1-7

[34] Taylor ZJ, Gurka R, Kopp GA, Liberzon A. Long-duration time-resolved PIV to study unsteady aerodynamics. IEEE Transactions on Instrumentation and Measurement. 2010;**59**:3262-3269

[35] Saigo T, Hamaguchi S. Shear viscosity of strongly coupled Yukawa systems. Physics of Plasmas. 2002;**9**: 1210-1216

[36] Donko Z, Goree J, Hartmann P, Kutasi K. Shear viscosity and Shear thinning in two-dimensional Yukawa liquids. Physical Review Letters. 2006; **96**:145003:1-4

Section 4

Vortices in Aerospace Engineering

Chapter 9

Wingtip Vortices of a Biomimetic Micro Air Vehicle

Rafael Bardera, Estela Barroso and Juan Carlos Matías

Abstract

Wingtip vortices are generated behind a wing that produces lift. They exhibit a circular pattern of spinning air that generates an additional drag force, the induced drag, reducing the aerodynamic performance of an aircraft. Moreover, the wingtip vortices can pose a hazard to airplane maneuvers, mainly in take-off and landing operations. This chapter describes a review of the lifting-line vortex theory applied to a biomimetic Micro Aerial Vehicle (MAV) with Zimmerman planform. Therefore, the horseshoe vortex model is deeply explained and the estimations of vortex velocity distribution, lift, and induced drag are obtained with this simple model. These results are compared with experimental data obtained from wind tunnel testing by using Particle Image Velocimetry (PIV). Finally, the vorticity maps in the wake of this MAV are obtained from PIV measurements.

Keywords: tip vortices, biomimetic, micro aerial vehicle, induced drag, vorticity

1. Introduction

The aeronautic industry has developed a growing interest in Unmanned Aerial Vehicles (UAVs). These vehicles have been designed for multiple missions where the human factor is not required. Therefore, in dangerous missions, unhealthy environments, or inaccessible areas, human accidents can be avoided. The UAVs can be distinguished into different categories according to their performance characteristics. In this context, the relevant design parameters are weight, manufacturing costs, and size. Mainly, the manufacturing costs have been the key point for that engineers and researchers could be focused on developing smaller vehicles in order to perform unmanned activities. This group of smaller vehicles is known as Micro Aerial Vehicles (MAVs) [1–3]. Their main features are low aspect ratio (AR) and low range operation. Research centers and universities have been able to investigate new designs of MAVs due to their low manufacturing costs and small size. This is the case of aerodynamic challenges posed in the work of Moschetta [4]. The MAV designs have taken into account the fixed-wing, coaxial, biplane, and tilt-body concept. Marek [5] performed experimental tests to determine the aerodynamic coefficients in six different types of platforms. The Zimmerman and elliptical planforms resulted in having the highest lift coefficient. Hence, Hassannalian and Abdelkefi [6] designed and manufactured a fixed-wing MAV based on the Zimmerman planform. Also, other authors designed the Dragonfly MAV using Zimmerman planform [7, 8].

The chapter will begin with a description of the biomimetic Micro Air Vehicle (MAV) [9, 10], then the horseshoe theory will be explained and applied to the studied vehicle. Consequently, the experimental facility, the Particle Image Velocimetry technique, and the description of the experimental tests will be defined. Then, the vorticity and several vortex models will be defined and applied to the experimental data obtained from the Wind Tunnel. At the end of the chapter, the formulation which relates the axial vorticity and the circulation will be presented and finally, the lift coefficient will be obtained.

2. Micro air vehicle geometry

The geometry of the studied Micro Air Vehicle (MAV) is based on Zimmerman planform and Eppler 61 airfoils for the wing configuration and Whitcomb II airfoils for the fuselage (see **Figure 1**).

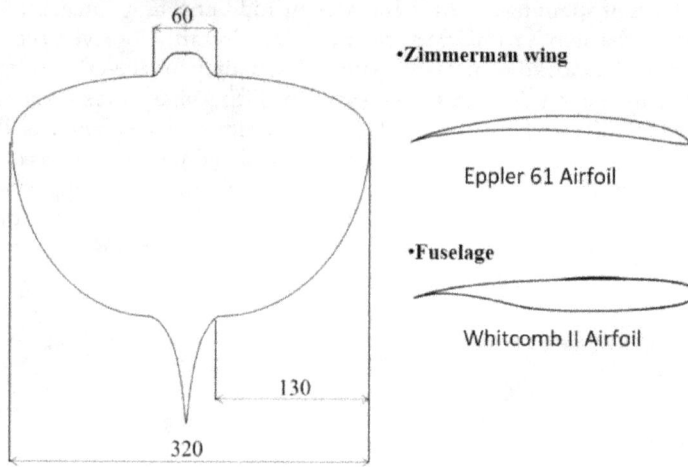

Figure 1.
Biomimetic MAV model (dimensions in mm).

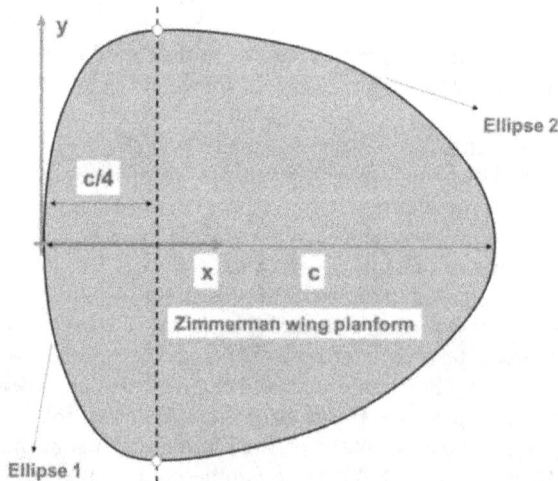

Figure 2.
Zimmerman planform.

Parameter	Value
Wing tip Chord c_t	0.025 m
Wing root Chord c_r	0.200 m
Wing taper ratio, λ	0.124
Aspect ratio, AR	2.500
Wingspan, b	0.320 m
Mean aerodynamic chord, CMA	0.141 m
Mean geometry chord, CMG	0.127 m
Wing reference area, S	0.040 m²
Dihedral angle, D_h	10°
Fuselage length, l	0.300 m
Fuselage width, d	0.060 m

Table 1.
MAV features.

The Zimmerman wing consists of two halves of ellipses connected at one point of reference which corresponds to 1/4 of the maximum wing root chord (c_r = 200 mm) for one half of the ellipse and 3/4 of the c_r for the other half of the ellipse. **Figure 2** shows the planform of the micro air vehicle and their dimensions. The rest of the geometrical features are shown in **Table 1**.

3. The horseshoe vortex: Biot-Savart law

In this section, a previous formulation of the wingtip vortex will be presented. The 3D wings can be modeled by vortex filaments. The horseshoe is the simplest mathematical model of potential flow to represent the aerodynamics of a wing aircraft. That consists of the bound vortex (vortex filament of the wing) and the trailing vortices formed by the semi-infinite filament vortex that represents the wingtips.

The horseshoe is a 3-D vortex that can be represented with an arbitrary shape according to the Helmholtz vortex theorems:

- The circulation Γ is constant along the vortex length.

- The vortex has to be extended to $\pm\infty$, form a closed-loop, or end at a solid boundary.

In this context, the velocity field of a 3-D vortex by applying the Biot-Savart Law is defined by the following expression Eq. (1), [11].

$$\vec{V}(x,y,z) = \frac{\Gamma}{4\pi} \int_{-\infty}^{+\infty} \frac{\vec{dl} \times \vec{r}}{\left|\vec{r}\right|^3} \tag{1}$$

where $\vec{r}$ is extended from the integration point to the field point and the arc length element $\vec{dl}$ points follow the direction of positive circulation.

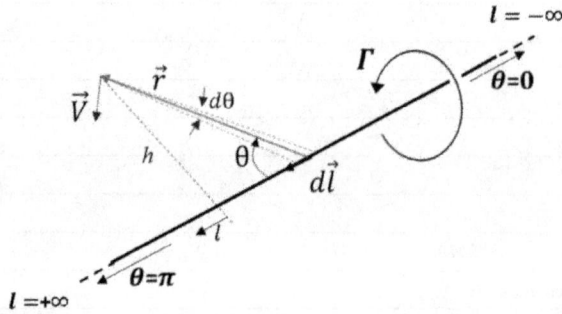

Figure 3.
Scheme of the straight-vortex.

Taking into account the straight vortex of **Figure 3**, h is defined as the nearest perpendicular distance from the vortex line and θ is the angle between the radius vector $\vec{r}$ and the vortex line (are defined in Eq. (4) and (5)).

$$r \equiv \left| \vec{r} \right| = \frac{h}{\sin\theta} \tag{2}$$

$$l = -\frac{h}{\tan\theta} \tag{3}$$

$$dl = \frac{h}{\sin^2\theta} \tag{4}$$

$$\vec{dl} \times \vec{r} = (dl\, r\, \sin\theta)\, \hat{\theta} \tag{5}$$

Now, the velocity field can be recalculated as Eq. (6):

$$\vec{V} = \frac{\Gamma}{4\pi h} \hat{\theta} \int_0^\pi \sin\theta\, d\theta = \frac{\Gamma}{2\pi h} \hat{\theta} \tag{6}$$

To reproduce the wingtip vortices of the studied MAV, a simple model based on the superposition of the freestream flow (U_∞) and a horseshoe vortex is described. The horseshoe vortex can be defined as the sum of three segments that can be seen in **Figure 4**: two free-trailing vortices at each tip of the wing (segment AB and segment CD) that are connected by a bound vortex spanning the wing (segment BC). As explained previously, the circulation Γ along the entire vortex line is constant, and the vortex lines have to extend downstream to infinity (see **Figure 3**). This potential solution is not very effective since the local lift to span is constant over the wingspan and in the real MAV model, the local lift is zero at the tip of the wings. A scheme of the horseshoe vortex model is defined in **Figure 4**.

The velocity field downstream of the wing in x = constant planes is similar to the potential solution generated by a horseshoe vortex except near the vortex axes. Now, to obtain the vertical velocity distribution of the potential vortex in our MAV, it is necessary to know the wing chord (c = 0.2 m), wingspan (b = 1.6c), and chord distance downstream of the trailing edge of the wing (x = 3c). Therefore, the following two non-dimensional variables (η and ζ) need to be defined (Eq. (7)):

$$\eta = \frac{x}{a} = 3.75;\ \zeta = \frac{y}{a} \tag{7}$$

where a is the semi-wingspan, $a = \frac{b}{2}$ (see **Figure 2**), η and ζ are the non-dimensional coordinates according to the x-axis and y-axis, respectively.

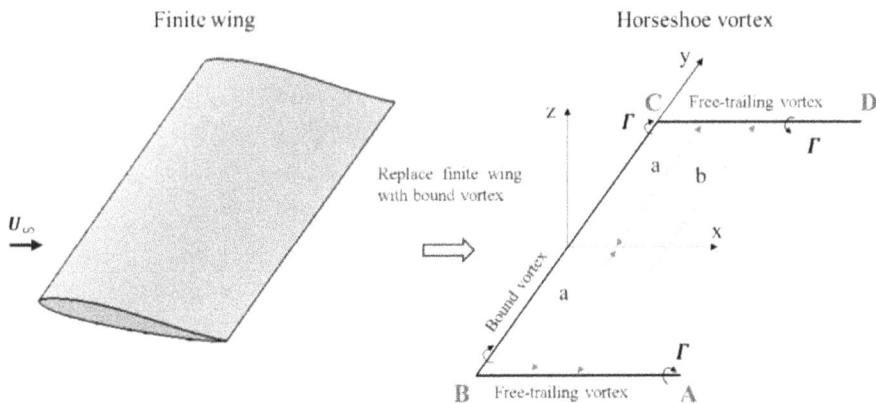

Figure 4.
Scheme of the horseshoe vortex model.

Then, the non-dimensional vertical velocity $\psi(\zeta)$ can be defined as Eq. (8):

$$\psi(\eta) = \frac{w(\zeta)}{\frac{\Gamma}{4\pi a}} \tag{8}$$

which presents a different formulation depending on the vortices defined in each of the segments (see **Figure 3**):

$$\psi_{AB}(\eta) = \frac{-1}{(\zeta+1)}\left[1 + \frac{\eta}{\sqrt{(\zeta+1)^2 + (\eta)^2}}\right] \tag{9}$$

$$\psi_{CD}(\eta) = \frac{1}{(\zeta-1)}\left[1 + \frac{\eta}{\sqrt{(\zeta-1)^2 + (\eta)^2}}\right] \tag{10}$$

$$\psi_{BC}(\eta) = \frac{-1}{\eta}\left[\left[\frac{\eta+1}{\sqrt{(\zeta+1)^2 + (\eta)^2}}\right] - \left[\frac{\eta-1}{\sqrt{(\zeta-1)^2 + (\eta)^2}}\right]\right] \tag{11}$$

Finally, the total non-dimensional vertical velocity is defined as the sum of the three velocities of the vortices (Eq. (12)):

$$\psi(\zeta) = \psi_{AB}(\zeta) + \psi_{BC}(\zeta) + \psi_{CD}(\zeta) \tag{12}$$

In the following **Figure 5**, the total non-dimensional vertical velocity distribution of this MAV is presented only for the section located at 3c downstream of the trailing edge of the wing and for the angle of attack of 10°.

To obtain a better understanding of the flow behavior of these vortices and how they interact between them, in **Figure 6** the non-dimensional vertical velocity only of the AB free-trailing vortex region is presented. The blue line shows the velocity distribution of the AB free-trailing vortex while the dashed red and black lines correspond to the velocity of the bound vortex (BC in **Figure 4**) and the CD free-trailing vortex, respectively. It is clearly noted that both vortices, bound vortex, and CD free-trailing vortex are not affecting the AB free trailing vortex since their

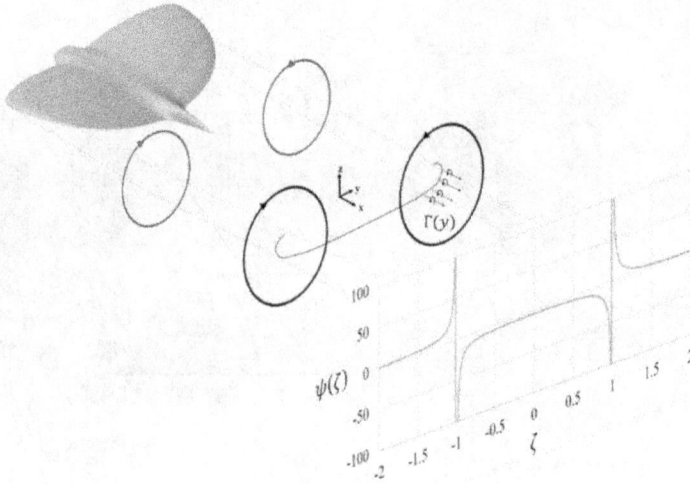

Figure 5.
The non-dimensional vertical velocity at 3c downstream of the trailing edge of the MAV wing.

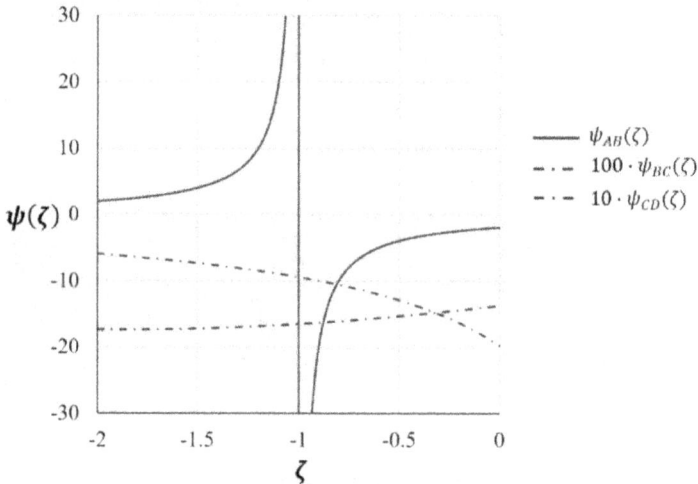

Figure 6.
The non-dimensional vertical velocity at 3c downstream of the trailing edge of the MAV wing.

velocity values are very small. As a consequence, in that region only the flow presence from the AB free-trailing vortex itself.

4. The experimental set-up

In this section, the experimental setup will be presented. All experimental tests were carried out in a Low-Speed Wind Tunnel at the Instituto Nacional de Técnica Aeroespacial (INTA) in Madrid (Spain). This wind tunnel has a closed circuit with an elliptical open test section of 6 m². The DC engine, which is situated at the opposite side of the test section, works at 420 V, allowing a maximum airflow speed of 60 m/s with a turbulence intensity lower than 0.5%. **Figure 7** shows the Low-Speed Wind Tunnel of INTA.

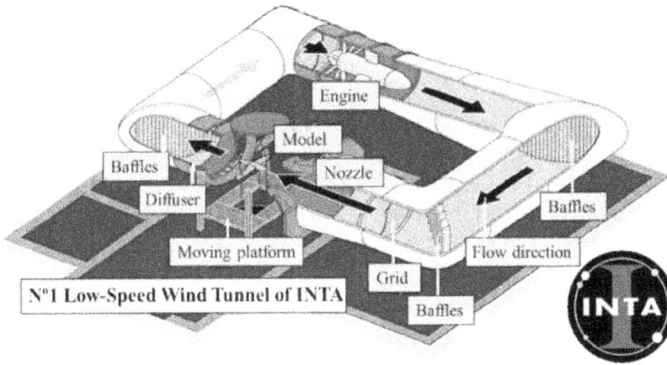

Figure 7.
Components of the low-speed wind tunnel of INTA.

The MAV model was tested with a freestream velocity of the wind tunnel of 10 m/s ($U_\infty = 10$ m/s), which results in a Reynolds number of 1.3×10^5 based on the wing root chord ($c_r = 0.20$ m). This analysis was performed for the cruise configuration (with an angle of attack equal to 0°). The experimental tests consisted in obtaining various transversal planes of the flow field at different sections downstream of the trailing edge of the wing.

The test experiments were carried out by using a full-scale model made in plastic material (PLA) by means of additive manufacturing and attached to a wood board by means of a streamlined support strut (see **Figure 8**). Only half of the model was studied due to its symmetry. Moreover, the MAV model and the wood board had to be painted in black in order to avoid reflections of the laser plane. The CCD camera was located behind the model (**Figure 8**), parallel to the flow stream of the wind tunnel.

The flow field velocity was determined by Particle Image Velocimetry (PIV). PIV is an advanced experimental technique that has the advantage of measuring the velocity field in a non-intrusive manner. This technique measures the velocity of the flow by analyzing flow images pairs [12]. For achieving this, PIV requires tracer particles that have to be seeded in the airflow. Olive oil was chosen for the

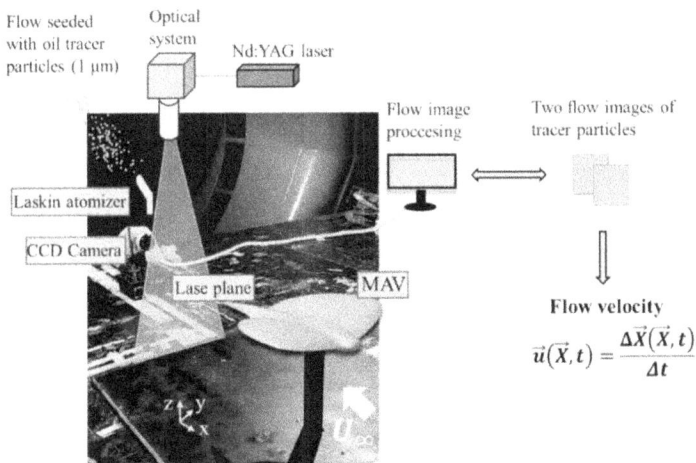

Figure 8.
Experimental setup.

generation of the tracer particles. A laser sheet has to be generated in order to go through the tracer particles and illuminate them. Two Neodymium-Yttrium Aluminum Garnet (Nd:YAG) lasers and an optical system were used for achieving this. The two lasers Nd:YAG has a pulse energy of 190 mJ within a time gap of 22 µs. The location of the tracer particles has to be recorded by a high-resolution camera with 2048×2048 pixels equipped with a lens Nikon Nikkor 50 mm 1:1.4D. A cross-correlation implemented via Fast Fourier Transform (FFT) is carried out over small image regions in order to obtain the averaged displacement of the tracer particles. The field of view (FOV) of the camera was 190×190 mm^2. The flow images are divided into interrogation window of 32×32 pixels. By using the Nyquist Sampling Criteria, these interrogation windows are overlapped by 50%. Moreover, the peak of correlation is adjusted to the subpixel accuracy by Gaussian approximation. A final post-processing task to remove spurious data and fill the empty vectors is needed. Therefore, a local mean filter based on a neighbor kernel window of 3×3 vectors was applied.

5. The vorticity in the wingtip wake

The vorticity is defined as the curl of the flow velocity, by the following expressions (Eq. (13) and Eq. (14)),

$$\vec{\omega} = \nabla \times \vec{V} \tag{13}$$

$$\vec{\omega} = \left(\frac{\partial w}{\partial y} - \frac{\partial v}{\partial z}\right)\vec{i} + \left(\frac{\partial u}{\partial z} - \frac{\partial w}{\partial x}\right)\vec{j} + \left(\frac{\partial v}{\partial x} - \frac{\partial u}{\partial y}\right)\vec{k} \tag{14}$$

The two-dimensional (2D y-z plane) streamwise vorticity $\omega_x = \xi$ can be determined from measured velocities by solving Eq. (15), which depends on the velocity spatial derivatives, as follows,

$$\xi = \frac{\partial w}{\partial y} - \frac{\partial v}{\partial z} = \left(\nabla \times \vec{V}\right) \cdot \vec{i} \tag{15}$$

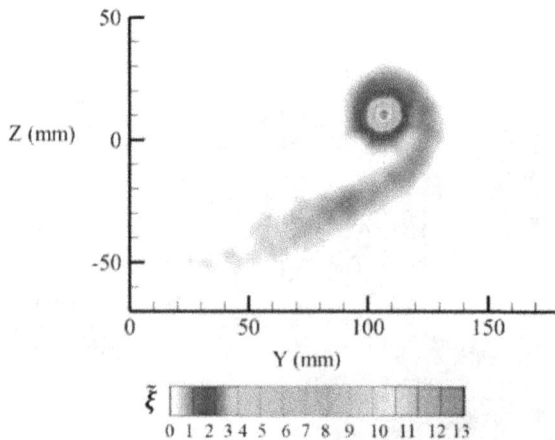

Figure 9.
Non-dimensional axial vorticity measured by PIV at 1.4 c downstream of the trailing edge of the model ($U_\infty = 10m/s, cruise : \alpha = \beta = 0°$).

The axial vorticity had to be obtained with central differencing in crossflow velocities. The non-dimensional form of axial vorticity component $\tilde{\xi}$ is given by the following expression (Eq. (16)):

$$\tilde{\xi} = \frac{\left(\frac{b}{2}\right) \cdot \xi}{U_\infty} \tag{16}$$

where b is the whole spanwise of the model and U_∞ is the frestream velocity.

Figure 9 shows the non-dimensional axial vorticity after taking PIV measurements in the wake downstream when the vehicle was flying in a cruise condition. It can be seen that the peak of maximum axial vorticity (red region) takes place at the wingtip, and from there it starts to decrease.

6. Circulation and vorticity

By analyzing the flow downstream of the aircraft model, this flow can be studied as the 2D wingtip wake and the vorticity is related to the velocity circulation from Stokes theorem by the following expression (Eq. (17)), [11].

$$\Gamma = \oint_C \vec{V} \cdot d\vec{l} = \iint \left(\nabla \times \vec{V}\right) \cdot \vec{n} \cdot dA \tag{17}$$

where C is a closed curve, $\vec{V}$ is the flow velocity on a small element defined on the closed curve, and dl is the differential length of that small element. As the plane streamwise is the 2D-yz plane, we have $\vec{\omega} = \xi \vec{i}$, and the unit normal vector $\vec{n} = \vec{i}$, then (Eq. (18)),

$$\Gamma = \oint_C \vec{V} \cdot d\vec{l} = \iint \vec{\omega} \cdot \vec{n} \cdot dA = \iint \xi \cdot dA \tag{18}$$

7. Evolution of the vorticity

The Navier-Stokes equations in vector form for an incompressible flow are given by,

$$\nabla \cdot \vec{V} = 0 \tag{19}$$

$$\frac{\partial \vec{V}}{\partial t} + \vec{V} \cdot \nabla \vec{V} = -\nabla \left(\frac{p}{\rho} + gz\right) + \nu \nabla^2 \vec{V} \tag{20}$$

The vorticity equation (Eq. (13)) is obtained by taking the curl of the Navier-Stokes equation, as follows,

$$\nabla \times \left(\nabla \cdot \vec{V}\right) = 0 \tag{21}$$

$$\nabla \times \left(\frac{\partial \vec{V}}{\partial t} + \left(\vec{V} \cdot \nabla\right) \vec{V} = -\nabla \left(\frac{p}{\rho} + gz\right) + \nu \nabla^2 \vec{V}\right) \tag{22}$$

By calculating each term, where the conservation of vorticity is Eq. (23),

$$\nabla \times \left(\frac{\partial \vec{V}}{\partial t}\right) = \frac{\partial \vec{\omega}}{\partial t} \tag{23}$$

$$\nabla \times \left(\left(\vec{V} \cdot \nabla\right)\vec{V}\right) = \left(\vec{V} \cdot \nabla\right)\vec{\omega} - \left(\vec{\omega} \cdot \nabla\right)\vec{V} \tag{24}$$

$$\nabla \times \left(-\nabla\left(\frac{p}{\rho} + gz\right)\right) = 0 \tag{25}$$

$$\nabla \times \left(\nu\nabla^2\vec{V}\right) = \nu\nabla^2\vec{\omega} \tag{26}$$

and finally, the vorticity equation is,

$$\frac{\partial \vec{\omega}}{\partial t} + \left(\vec{V} \cdot \nabla\right)\vec{\omega} = \left(\vec{\omega} \cdot \nabla\right)\vec{V} + \nu\nabla^2\vec{\omega} \tag{27}$$

The law of vorticity evolution is a convective vector diffusion equation given by the following expression,

$$\frac{D\vec{\omega}}{Dt} = \left(\vec{\omega} \cdot \nabla\right)\vec{V} + \nu\nabla^2\vec{\omega} \tag{28}$$

The viscous term $(\nu\nabla^2\vec{\omega})$ causes the vortex to diffuse through the fluid flow. By using index notation, the vorticity equation for 3D flow is given by,

$$\frac{\partial \omega_i}{\partial t} + u_j\frac{\partial \omega_i}{\partial x_j} = \omega_j\frac{\partial u_i}{\partial x_j} + \nu\frac{\partial^2 \omega_i}{\partial x_k \partial x_k} \tag{29}$$

For a 2D flow, the stretching term is absent, and the corresponding equation is,

$$\frac{\partial \omega_i}{\partial t} + u_j\frac{\partial \omega_i}{\partial x_j} = \nu\frac{\partial^2 \omega_i}{\partial x_k \partial x_k} \tag{30}$$

Equivalently, in vector form, for a 2D flow we have the velocity is perpendicular to the vorticity, so $\vec{V} \cdot \vec{\omega} = 0$. The velocity is $\vec{V} = (0, V, W)$ and vorticity $\vec{\omega} = (\omega_x, 0, 0)$, so that,

$$\vec{\omega} \cdot \nabla\vec{V} = 0 \tag{31}$$

$$\frac{D\vec{\omega}}{Dt} = \frac{\partial \vec{\omega}}{\partial t} + \left(\vec{V} \cdot \nabla\right)\vec{\omega} = \nu\nabla^2\vec{\omega} \tag{32}$$

where the operator $\frac{D}{Dt} = \frac{\partial}{\partial t} + \left(\vec{V} \cdot \nabla\right)$ is the material derivative and it describes the evolution along the flow lines.

8. Decay of wingtip vortices

The study of the decay of wingtip vortices under the assumption of 2D flow with $\omega_y = \omega_z = 0$, velocity $V_x = 0$ and $\partial/\partial x = 0$, can be performed by the 2D viscous diffusion of vorticity equation, given by,

$$\frac{\partial \vec{\omega}}{\partial t} = \nu \nabla^2 \vec{\omega} \tag{33}$$

$$\frac{\partial \omega}{\partial t} = \nu \cdot \Delta \omega \tag{34}$$

Where $\omega = \omega_x$ and Δ is the Laplacian operator.
Assuming axisymmetric flow, in cylindrical coordinates,

$$\frac{\partial \omega}{\partial t} = \frac{\nu}{r} \cdot \frac{\partial}{\partial r}\left(r \frac{\partial \omega}{\partial r}\right) \tag{35}$$

The initial vorticity for the study of decay point vortex in an unbounded domain is given by a 2D delta function in the plane yz,

$$\omega\left(\vec{x}, t = 0\right) = \Gamma_0 \delta(y)\delta(z) \tag{36}$$

Introducing the dimensionless similarity variable [13],

$$\epsilon = \frac{r}{\sqrt{\nu t}} \tag{37}$$

and the nondimensional combination $\omega \nu t / \Gamma_0$ can be expressed as an unknown function g of the variable ϵ, defined as

$$\omega \nu t / \Gamma_0 = g(\epsilon) \tag{38}$$

So that,

$$\omega = \frac{\Gamma_0}{\nu t} g(\epsilon) = f(t) g(\epsilon) \tag{39}$$

Calculating the derivatives quantities from the earlier equation,

$$\frac{\partial \omega}{\partial t} = \frac{\partial f(t)}{\partial t} g(\epsilon) + f(t) \frac{\partial g(\epsilon)}{\partial t} = -\frac{\Gamma_0}{\nu t} \frac{1}{t} g(\epsilon) + f(t) \frac{dg(\epsilon)}{d\epsilon} \frac{\partial \epsilon}{\partial t} \tag{40}$$

$$\frac{\partial \omega}{\partial t} = -f(t) \frac{1}{t} g(\epsilon) - f(t) \frac{\epsilon}{2t} \frac{dg(\epsilon)}{d\epsilon} = -f(t) \frac{1}{t}(g + \epsilon g'/2) \tag{41}$$

On the other hand,

$$\frac{\partial \omega}{\partial r} = f(t) \frac{\partial g(\epsilon)}{\partial r} = f(t) \left(\frac{\partial \epsilon}{\partial r} \frac{dg(\epsilon)}{d\epsilon}\right) = f(t) \left(\frac{\epsilon}{r} \frac{dg(\epsilon)}{d\epsilon}\right) \tag{42}$$

$$\frac{\partial \omega}{\partial r} = f(t) \frac{\partial g(\epsilon)}{\partial r} = f(t) \left(\frac{\partial \epsilon}{\partial r} \frac{dg(\epsilon)}{d\epsilon}\right) = f(t) \left(\frac{\epsilon}{r} \frac{dg(\epsilon)}{d\epsilon}\right) \tag{43}$$

$$\frac{\partial}{\partial r}\left(r \frac{\partial \omega}{\partial r}\right) = \frac{\partial \epsilon}{\partial r} \frac{d}{d\epsilon}\left(f(t) \, \epsilon \, \frac{dg(\epsilon)}{d\epsilon}\right) = \frac{\epsilon}{r} f(t) \frac{d}{d\epsilon}\left(\epsilon \, \frac{dg(\epsilon)}{d\epsilon}\right) \tag{44}$$

And substituting in (35), the following expression is derived,

$$2(\epsilon g')' + \epsilon^2 g' + 2g\epsilon = 0 \tag{45}$$

Where ' denotes a derivative respect to, and finally, the equation is integrated

$$g(\epsilon) = A \, \exp\left(\frac{-\epsilon^2}{4}\right) \tag{46}$$

The condition of the flow circulation is equal to Γ_0 at any time, gives,

$$\Gamma_0 = \int_0^\infty \omega 2\pi r \, dr = 4\pi A \Gamma_0 \tag{47}$$

so that $A = 1/4\pi$, and the solution of the $g(\epsilon)$ function is,

$$g(\epsilon) = \frac{1}{4\pi} \, \exp\left(\frac{-r^2}{4\nu t}\right) \tag{48}$$

Finally, the solution of vorticity is given by the axisymmetric Lamb-Osteen vortex by,

$$\omega = \frac{\Gamma_0}{4\pi\nu t} \, \exp\left(\frac{-r^2}{4\nu t}\right) \tag{49}$$

The swirl velocity is,

$$V_\theta = \frac{\Gamma_0}{2\pi r} \left(1 - \exp\frac{-r^2}{4\nu t}\right) \tag{50}$$

and the circulation is,

$$\Gamma = \Gamma_0 \left(1 - \exp\frac{-r^2}{4\nu t}\right) \tag{51}$$

The swirl velocity can be rewritten as,

$$V_\theta = \frac{\Gamma_0}{2\pi r} \left(1 - \exp\left(-1.2526 \left(r/r_c\right)^2\right)\right) \tag{52}$$

where r_c is the core radius, defined as the distance from the vortex center to the point with the higher swirl velocity, and given by,

$$r_c = 2.24\sqrt{\nu t} \tag{53}$$

9. Analysis of vortex models and experimental data

The velocity components which define a 2-D vortex are typically the swirl velocity V_θ, the axial velocity V_z and the radial V_r velocity. The last two components usually are neglected in many applications as they are very small compared to swirl velocity, and are defined as follows,

$$V_\theta = \frac{\Gamma}{2\pi r} \tag{54}$$

$$V_r = 0 \tag{55}$$

$$V_z = 0 \tag{56}$$

Several tip vortex models are usually studied, but this chapter is only focused on some of them, displayed in **Figure 10**. The first method is the Rankine vortex model, being the simplest formulation with a finite core. Therefore, the vortex is divided into two parts: the viscous core and the potential vortex. The viscous core is rotating as a solid body near the vortex center while the potential vortex remains away from the vortex center. The velocity in the potential vortex is decreasing hyperbolically with the radial coordinate [14, 15]. Therefore, the following expressions represent the swirl velocity distribution V_θ in the tip vortex,

$$V_\theta(\tilde{r}) = \left(\frac{\Gamma}{2\pi r_c}\right) \bullet \tilde{r} \ 0 \leq \tilde{r} \leq 1 \tag{57}$$

$$V_\theta = \frac{\Gamma}{2\pi r} \ \tilde{r} > 1 \tag{58}$$

Where r_c is the viscous core radius and $\tilde{r} = \frac{r}{r_c}$ is the non-dimensional radial coordinate.

The second vortex model is the Lamb-Oseen vortex which is a simplified solution of one-dimensional Navier-Stokes equations for laminar flow which is defined by the following expression,

$$V_\theta(\tilde{r}) = \left(\frac{\Gamma}{2\pi r}\right) \bullet \left[1 - e^{-\alpha(\tilde{r})^2}\right] \tag{59}$$

where $\alpha = 1.2526$ is the Oseen parameter.

An alternative tip vortex formulation is given by Vatistas in Ref. [15]. This method is based on a group of desingularized algebraic swirl velocity profiles for vortices which present continuous distributions of flow quantities. The swirl velocity is defined by,

$$V_\theta(\tilde{r}) = \frac{\Gamma}{2\pi r_c} \bullet \frac{\tilde{r}}{\left(1 + \tilde{r}^{2n}\right)^{1/n}} \tag{60}$$

where n is an integer.

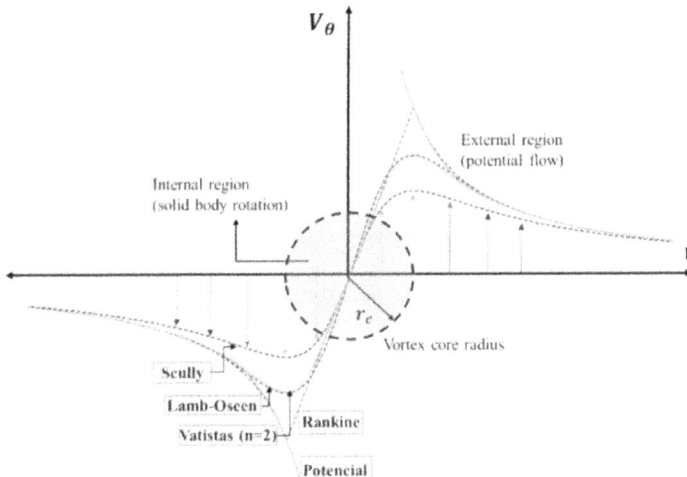

Figure 10.
Swirl velocity distribution inside a tip vortex was obtained by several tip vortex models.

The Scully vortex model is the previous formulation when the integer is $n = 1$, and it is defined as,

$$V_\theta(\tilde{r}) = \left(\frac{\Gamma}{2\pi r_c}\right) \cdot \frac{\tilde{r}}{(1+\tilde{r}^2)} \qquad (61)$$

when the integer is $n = 2$, the swirl velocity of the vortex formulation is,

$$V_\theta(\tilde{r}) = \left(\frac{\Gamma}{2\pi r_c}\right) \cdot \frac{\tilde{r}}{\sqrt{1+\tilde{r}^4}} \qquad (62)$$

It is important to notice that when the integer $n \to \infty$, the swirl velocity distribution corresponds to the Rankine method.

Figure 11 shows the flow field velocity in a normal section to the flow located at 3 chords downstream of the MAV model. The 2d vortex can be observed clearly and the color scale indicates that the velocity is increasing near the center of the vortex.

It is possible to obtain a better visualization of the flow field distribution by looking at **Figure 12**. This PIV map is obtained for the angle of attack of 10°. The plotted streamlines reveal the location of the vortex center (places at Y = Z = 0 mm), the region of the vortex core (yellow region), and the external region (green area).

Extracting the data value of the swirl velocity as measured by the PIV technique we can obtain **Figure 13** when the experimental data are plotted with curves of theoretical vortex models. The blue scatter dots which its trend is approached by a 6th-degree polynomial (red continue line).

Also, the distributions of the swirl velocity obtained by the theoretical vortex models as Rankine, Lamb-Oseen, and Scully are represented in **Figure 11**.

The analysis of this graph shows the wingtip vortex method which presents the most accurate fit to the MAV is obtained with the tip vortex model of Scully. Subsequently, there is a deviation between the two approaches (experimental data and Scully) which depends on the distance from the vortex center. The ratio between both of them is assessed by the parameter $k(r)$ defined as

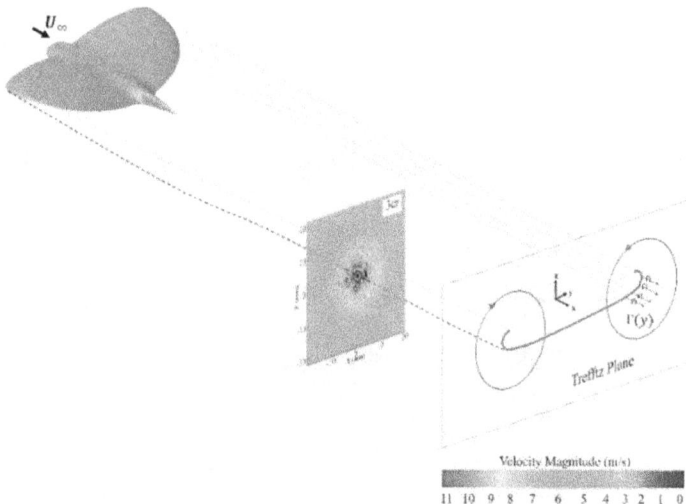

Figure 11.
Wingtip vortex in the MAV wake at 3c.

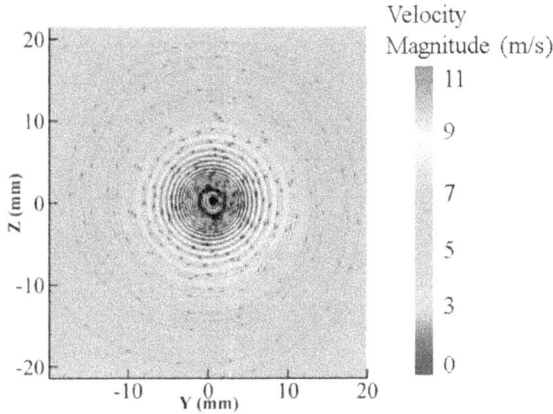

Figure 12.
Velocity distribution at 3c.

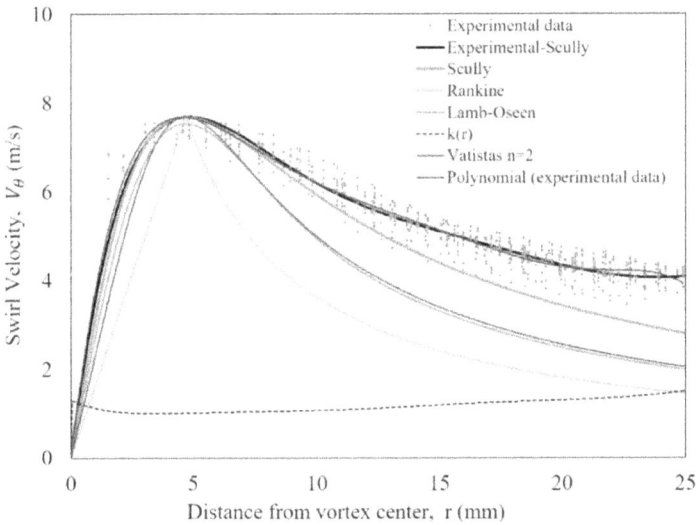

Figure 13.
Experimental data and theoretical vortex models.

$$k(r) = \frac{(V_\theta)_{polynomial}}{(V_\theta)_{Scully}} \quad (63)$$

where $(V_\theta)_{polynomial}$ and $(V_\theta)_{Scully}$ are the distributions of swirl velocity obtained in the test experiments and by the theoretical model proposed by Scully, respectively.

Finally, the distribution of experimental swirl velocity is fitted to the Scully model by the function called $(V_\theta)_{experimental-Scully}$ defined as,

$$(V_\theta)_{experimental-Scully} = k(r) \cdot (V_\theta)_{Scully} \quad (64)$$

10. Lift coefficient

The lift of an airfoil can be determined by the Kutta-Joukowski theorem [11] relating the velocity and the circulation, as follows,

Parameters	MAV model
Location	x = 3 chords
$\alpha(°)$	0
$\beta(°)$	0
$U_\infty (m/s)$	10
r_c (mm)	4.70
$V_{\theta max}$ (m/s)	7.69
$\Gamma (m^2/s)$	0.45
C_L	0.72

Table 2.
Results of the tip vortex analysis in the wake of MAV.

$$L' = \rho U_\infty \Gamma \tag{65}$$

By applying the earlier formulation, the total lift of the wing L can be obtained from the following expression

$$L = (\rho U_\infty \Gamma) \bullet b \tag{66}$$

where b is the wingspan.

The lift coefficient C_L is obtained by dividing the lift by $q_\infty S_{ref}$,

$$C_L = \frac{(\rho U_\infty \Gamma) \bullet b}{q_\infty S_{ref}} \tag{67}$$

where q_∞ is the dynamic pressure ($q_\infty = 1/2\rho U_\infty^2$) and S_{ref} is the reference wing surface.

Table 2 shows the values of the main parameters obtained from the tip vortex analysis, including the lift coefficient, C_L.

11. Conclusions

Wingtip vortices generated behind an aircraft wing affect the aerodynamic performance of the aircraft while endangering take-off and landing maneuvers of the subsequent aircraft.

In this chapter, it is reviewed the theoretical background of the horseshoe vortex and several vortex models applied to a Biomimetic Micro Air Vehicle (MAV) with Zimmerman planform. The formulation of the vorticity in the wingtip wake of the MAV model has been presented as well as the expression which relates the axial vorticity and the circulation.

All experimental tests have been carried out in the Low-Speed Wind Tunnel of the Instituto Nacional de Técnica Aeroespacial (INTA) with a full-scale MAV model. Particle Image Velocimetry has been used to obtain the transversal flow field at 3 chords downstream of the trailing edge of the MAV model. The swirl velocity distribution according to the horseshoe vortex model and several vortex models (Rankine, Lamb-Oseen, Scully, and Vatistas) is plotted. The experimental results have shown that the Scully vortex has the most similar behavior to the MAV wing vortex. The distribution of the transversal velocity as well as the axial vorticity for the section at 3 chords are presented by PIV maps. Finally, the lift coefficient by using the Kutta-Joukowski theorem is obtained.

Acknowledgements

This investigation was funded by the Spanish Ministry of Defense under the program "464A 64 1999 14205005 Termofluidodinámica" with internal code IGB 99001 of Instituto Nacional de Técnica Aeroespacial "Esteban Terradas" (INTA-National Institute for Aerospace Technology of Spain).

Conflict of interest

The authors declare that they have not known existing or potential Conflicts of Interest, including financial or personal factors, as well as any relationship which could influence their scientific work.

Author details

Rafael Bardera*, Estela Barroso and Juan Carlos Matías
National Institute for Aerospace Technology, Madrid, Spain

*Address all correspondence to: barderar@inta.es

IntechOpen

References

[1] Muller. Aerodynamic Measurements at Low Reynolds Numbers for Fixed Wing Micro-Air Vehicles. Notre Dame, IN, USA: Hessert Center for Aerospace Research; 1999

[2] Torres. Low-aspect-ratio wing aerodynamics at low reynolds numbers. AIAA Journal. 2004;**42**:865-873

[3] Mizoguchi M, Kajikawa Y, Itoh H. Aerodynamic characteristics of low-aspect-ratio wings with various aspect ratios in low reynolds number flows. Transactions of the Japan Society For Aeronautical and Space Sciences. 2016; **59**:56-63

[4] Moschetta JL. The aerodynamics of micro air vehicle: Technical challenges and scientific issues. International Journal of Engineering Systems Modelling and Simulation. 2014;**6**(3/4):134-148

[5] Marek PL. "Design, optimization and flight testing of a micro air vehicle" Doctoral dissertation. Glasgow, Scotland: University of Glasgow; 2008

[6] Hassanalian M, Abdelkefi A. Design and manufacture of a fixed wing MAV with Zimmerman planform. AIAA Sci Tech. AIAA 2016-1743. 54th AIAA Aerospace Sciences Meeting, 4-8 January 2016, San Diego, California, USA. 2016. DOI: 10.2514/6.2016-1743

[7] Stanford B, Sytsma M, Albertani R, Viieru D, Shyy W, Ifju P. Static aeroelastic model validation of membrane micro air vehicle wings. AIAA Journal. 2007;**45**(12):2828-2837

[8] Flake J, Frischknecht B, Hansen S, Knoebel N, Ostler J, Tuley B. "Development of the Stableyes Unmanned Air Vehicle", 8th International Micro Air Vehicle Competition. Tucson, AZ: The University of Arizona; 2004. pp. 1-10

[9] Barcala-Montejano MA, Rodríguez-Sevillano A, Crespo-Moreno J, Bardera Mora R, Silva-González AJ. Optimized performance of a morphing micro air vehicle. Unmanned Aircraft Systems (ICUAS), 2015 International Conference. IEEE Journal of Intelligent Material Systems and Structures. Denver Marriott Tech Center, Denver, Colorado, USA: IEEE; June 9-12, 2015. pp. 794-800. DOI: 10.1109/ICUAS.2015.7152363

[10] Barcala-Montejano MA, Rodríguez-Sevillano A, Bardera-Mora R, García-Ramirez J, Leon-Calero M, Nova-Trigueros J. "Development of a morphing wing in a micro air vehicle" SMART. 8th Conference on Smart Structures and Materials - ECCOMAS, Madrid, Spain. 2017

[11] Anderson JD. Fundamentals of Aerodynamics. 2nd ed. USA: McGraw-Hill; 1991

[12] Prasad AK. Particle image velocimetry. Current Science. 2000; **79**(1):51-60

[13] Pozrikidis C. Introduction to Theoretical and Computational Fluid Dynamics. second ed. USA: Oxford University Press (OUP); 1997. p. 268

[14] Djojodihardjo H. Analysis and visualization studies of near field aircraft trailing vortices for passive wake alleviation. Novosibirsk, Russia: The 13th Asian symposium on visualization, June 22–26; 2015

[15] Mahendra JJ, Gordon L. Generalized viscous vortex model for application to free-vortex wake and aeroacoustic calculations. University of Maryland, College Park, Maryland: Alfred Gessow Rotorcraft Center, Department of Aerospace Engineering, Glenn L. Martin Institute of Technology; 2002

www.ingramcontent.com/pod-product-compliance
Lightning Source LLC
Chambersburg PA
CBHW081558190326
41458CB00015B/5649